W0080229

# Springer Earth System Sciences

**Series editors**

Philippe Blondel, Bath, UK
Eric Guilyardi, Paris, France
Jorge Rabassa, Ushuaia, Argentina
Clive Horwood, Chichester, UK

More information about this series at http://www.springer.com/series/10178

Patricia Eugenia Zalba · Martín Eduardo Morosi
María Susana Conconi

# Gondwana Industrial Clays

Tandilia System, Argentina: Geology
and Applications

 Springer

Patricia Eugenia Zalba
Comisión de Investigaciones Científicas de
la Provincia de Buenos Aires, Centro de
Tecnología de Recursos Minerales y
Cerámica
(CETMIC-CICPBA-CONICET)
Manuel B. Gonnet
Argentina

and

School of Geosciences
Monash University
Melbourne
Australia

Martín Eduardo Morosi
Comisión de Investigaciones Científicas de
la Provincia de Buenos Aires, Centro de
Tecnología de Recursos Minerales y
Cerámica
(CETMIC-CICPBA-CONICET)
Manuel B. Gonnet
Argentina

and

Facultad de Ciencias Naturales y Museo de
La Plata (UNLP)
La Plata
Argentina

María Susana Conconi
Comisión de Investigaciones Científicas de
la Provincia de Buenos Aires, Centro de
Tecnología de Recursos Minerales y
Cerámica
(CETMIC-CICPBA-CONICET)
Manuel B. Gonnet
Argentina

and

Facultad de Ingeniería de La Plata (UNLP)
La Plata
Argentina

ISSN 2197-9596                    ISSN 2197-960X  (electronic)
Springer Earth System Sciences
ISBN 978-3-319-39455-8            ISBN 978-3-319-39457-2   (eBook)
DOI 10.1007/978-3-319-39457-2

Library of Congress Control Number: 2016947209

© Springer International Publishing Switzerland 2016
This work is subject to copyright. All rights are reserved by the Publisher, whether the whole or part
of the material is concerned, specifically the rights of translation, reprinting, reuse of illustrations,
recitation, broadcasting, reproduction on microfilms or in any other physical way, and transmission
or information storage and retrieval, electronic adaptation, computer software, or by similar or dissimilar
methodology now known or hereafter developed.
The use of general descriptive names, registered names, trademarks, service marks, etc. in this
publication does not imply, even in the absence of a specific statement, that such names are exempt
from the relevant protective laws and regulations and therefore free for general use.
The publisher, the authors and the editors are safe to assume that the advice and information in this
book are believed to be true and accurate at the date of publication. Neither the publisher nor the
authors or the editors give a warranty, express or implied, with respect to the material contained
herein or for any errors or omissions that may have been made.

Printed on acid-free paper

This Springer imprint is published by Springer Nature
The registered company is Springer International Publishing AG Switzerland

# Preface

The information provided herein is the result of research conducted specifically for this work; from own research projects sponsored by the Comisión de Investigaciones Científicas de la provincia de Buenos Aires (CICPBA); Facultad de Ciencias Naturales y Museo de La Plata, and from cooperation with interdisciplinary research groups, mainly with researchers from the Consejo Nacional de Investigaciones Científicas y Técnicas (CONICET), through the Centro de Tecnología de Recursos Minerales y Cerámica (CETMIC), the Universidad de Buenos Aires (UBA), the Centro de Investigaciones Geológicas (CIG), international cooperation projects with the University of Poitiers, France, and from the international and local geological literature of researchers who worked on the study area, available at the end of each chapter.

Photographs, stratigraphic sections, drawings, chemical and technological analyses, mineralogical analyses by optical and scanning electron microscopy of our own, as well as some obtained from the literature, are offered. Besides, a glossary has been included to facilitate understanding of geological and technological terms to the reader.

The book is organized after a synthesis of the geology and stratigraphy of Tandilia, being the clay deposits of economic importance arranged in accordance to the counties and the sectors where they outcrop as follows:

| | |
|---|---|
| Azul County | Chillar Sector |
| Benito Juárez County | El Ferrugo and Constante 10-El Cañón Sector |
| | Villa Cacique Sector |
| | La Juanita Sector |
| | Cuchilla de Las Aguilas-Sierra de LaTinta Sector |
| Lobería County | San Manuel Sector |
| Olavarría County | Sierras Bayas Sector |
| General Pueyrredón, Balcarce and Necochea counties | Mar del Plata (Chapadmalal)-Balcarce-Necochea Sector |

Researchers immersed in the study of geological problems often disagree on the interpretation of occurring processes. But this is normal and specifically occurs in Tandilia because of the complexity and overlapping of physico-chemical and structural phenomena whose interpretation is even more difficult because of the antiquity of the deposits. As Prof. Walter Keller (†), eminent researcher on clay mineralogy of the University of Missouri, USA, said once: "if several expert scientists meet in an outcrop and their interpretation of the geology of the area was requested, one would obtain as many versions as geologists are involved." In this precise case, numerous and accurate observations, and a scientific demonstration of the processes that took place, shall cooperate in the development of the knowledge of this complex geology. That is why working as a team, confronting and discussing results among colleagues, in national and international forums, will bring closer hypothesis to reality, but this approximation will never be definitive. There will always be new discoveries, new theories, new techniques, which will allow us to evolve in the right direction, toward a greater conviction. And this is the challenge of geology.

We are deeply indebted to Springer for publishing our book which will reach a wider audience via its English version. As the book is a review of the research carried out for more than 30 years by the authors parts of it were only published in Spanish. Our special thanks to Dr. Jorge Rabassa, Springer editor, for his encouragment and useful aid when performing the organization and content of the manuscript.

We are grateful to the Comisión de Investigaciones Científicas de la Provincia de Buenos Aires for its financial support and to his past President Ing. José Rodríguez Silveira for his interest and permanent encouragement for the publication of this book. Also we would like to thank to the Centro de Tecnología de Recursos Minerales y Cerámica and to the Facultad de Ciencias Naturales y Museo de La Plata (UNLP) for their permamment support to our research. Our thanks to Lic. Nelson Coriale for useful information on the quarries. Our special thanks to Prof. Susana Zalba for invaluable help in improving translation to English. Finally, we thank permission for the access to the quarries to: Loma Negra CIASA, Cementos Avellaneda SA, Cerro Negro SA, Palmar Mar del Plata SA, Transportes Arcamin SA, Polcecal SA, Cruz Omar Pavone SRL, Yaraví, Delia Isabel Etchegoin, Cefas SA, Vial SA,Tagliorette, Rottemberger, Della Maggiora, and H. Foster.

Manuel B. Gonnet, Argentina                              Patricia Eugenia Zalba
Melbourne, Australia                                     Martín Eduardo Morosi
                                                         María Susana Conconi

# Contents

# Chapter 1
# Overview

**Abstract** This book is directed to industrials, students, and researchers who work with clays as raw material and who are interested in the behavior of clays in the final product. In fact, it is directed to everyone interested in the problematic of clay minerals. This chapter offers a synthesis of the history of the research carried out through more than 153 years on the Tandilia System, also known as Sierras Septentrionales de Buenos Aires, beginning with the pioneer work of Heusser and Claraz (1863). The residual and sedimentary deposits are organized in five productive counties and eight mining sectors of the province, where economically important Neoproterozoic and Lower Paleozoic clay reserves are found. Also, a general view of the regional geology and stratigraphy, a geologic map of Tandilia (taken from Iñíguez et al. 1989), with the main clay deposits of industrial importance and a stratigraphic scheme for different areas, some of them reviewed, are shown. The Tandilia basin is small today in relation with its original dimension when the Gondwana continent was intact. Nowadays, most of its sediments remain in South Africa, after the breakup of the continents which began in the Jurassic period. In spite of being sediments from almost 700 Ma of age, most of them have preserved their physico-chemical properties unchanged. Stratigraphic sections, mineralogical, chemical, differential thermal, X-ray diffraction analyses, optical and scanning electron microscopy, as well as technological analyses from our research of more than 30 years in collaborative working, and from the geological literature, are offered. A Glossary of current terms used in the vocabulary is included at the end of the book and specific references are available at the end of each chapter.

**Keywords** Tandilia System · Geological history · Stratigraphic scheme · Geologic map · Industrial clays

© Springer International Publishing Switzerland 2016                    1
P.E. Zalba et al., *Gondwana Industrial Clays*,
Springer Earth System Sciences, DOI 10.1007/978-3-319-39457-2_1

## 1.1 Introduction

As it is well known in Argentina, the Province of Buenos Aires is the main producer of the country in industrial rocks (granitoids, clays, limestones, dolostones, quartzites, and the last one known as "Mar del Plata stone") in terms of volume of exploitation.

The largest deposits of clays known so far distributed in the Tandilia System (Nágera 1940), also known as Sierras Septentrionales de Buenos Aires, are mainly around the cities of Olavarría and Tandil (Fig. 1.1). They are found in Precambrian and Eopaleozoic sedimentary sequences and overlying crystalline basement rocks (igneous and metamorphic). In the latter, in some cases economically exploitable residual deposits have been formed by alteration processes.

Crystalline basement rocks (granites, migmatites, gneisses, tonalites, amphibolites, etc.) are covered by sedimentary rocks formed by clastic deposits (derived from the destruction of preexisting rocks) of different particle size and mineralogical composition (conglomerates, breccias; sandstones: quartzites; shales: siltstones and claystones), chemical deposits (limestones, marls), and diagenetic (dolostones) some of which are of great importance, mainly in the construction industry.

The sedimentary deposits, which are part of different geological formations, and of different age, are easily identifiable and are separated by paleosurfaces (which represent unconformities) whose recognition has helped to explain their different

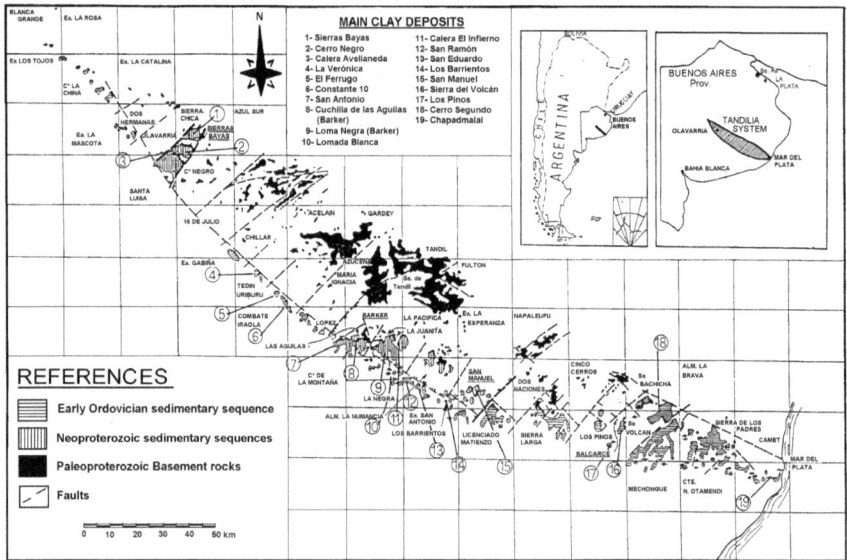

**Fig. 1.1** Geologic map of Tandilia System showing the location of main clay deposits found in the Tandilia System. Taken from Iñíguez et al. (1989)

ages, the areal distribution and, mainly, has contributed to the geological correlation of different outcrops of Tandilia, or have been identified by drilling and by the opening of quarries.

The detailed knowledge of the sedimentary deposits is important for several reasons. First for their economic importance, also for their strategic location with respect to major urban centers and, finally, for the quality of the raw materials they contain. Moreover, these deposits are some of the oldest known on Earth (Proterozoic: since 2000 million years; Neoproterozoic: 760–600 million years; and Lower Paleozoic: 500–435 million years) whose physico-chemical properties have been preserved almost intact for their use in industry.

The geology of Tandilia is not simple; epeirogenic movements that occurred in the Precambrian led to the differential rise of blocks of sedimentary deposits of the original basin and of the underlying crystalline basement rocks as well. This phenomenon conditioned the mining process, affecting the lateral continuity of the deposits: the uplifted blocks experimented erosion and those down-lifted were hidden at variable depths. Also this phenomenon is responsible for the difficulties in recognizing one particular deposit in different areas. Therefore, in interpreting the geology of a specific area, the geology of the whole range must, inevitably, be known.

## 1.2 A Bit of History

Today, the Tandilia basin, province of Buenos Aires, is small in relation to its original dimensions, most of which was restricted to South Africa, after the separation of the continents, which began in the Jurassic period (approximately 183 million years ago). For this reason, Tandilia represents only a small part of one great puzzle that cannot be recomposed in South America but by stratigraphical correlations with the African continent, and this has been one major problem when trying to reconstruct its geological history. Much work has been carried out in this regard through numerous studies (geological, stratigraphic, biostratigraphic, sedimentological, paleoenvironmental, mineralogical, structural, geochronological and, more recently, diagenetic) that began with the pioneer work of Heusser and Claraz (1863) in Tandilia. During the 1970s detail studies began to develop and this represented an important achievement related to new interpretations and reformulation of the classical stratigraphic scheme laid out for more than 150 years.

The change was not fortuitous. The transcendental shift in the approach of sedimentology, occurred around 1950 all over the world, settled new bases for detailed studies. Regionally, in Argentina, in the 1970s the geological review of Tandilia began to be carried out in a 1:50,000 scale, known as Plan Tandilia, under the direction of Prof. Dr. Adrián Mario Iñíguez, sponsored by the Comisión the Investigaciones Científicas de la provincia de Buenos Aires, and by the Centro de Investigaciones Geológicas, with the participation of the Centro de Tecnología de Recursos Minerales y Cerámica. The geological charts carried out were presented in full in the Primeras Jornadas Geológicas Bonaerenses, carried out in the city of Tandil, in 1985. This map represented an extraordinary advance in the geological knowledge of this amazing mountain range. However, unfortunately, it was never published in full.

From the works of Holmberg (1972), Amos et al. (1972), Dalla Salda et al. (1972), Dalla Salda and Iñíguez (1979), new advances in the stratigraphy of Tandilia materialized when the authors recognized that there was not a single sedimentary sequence (also named series, group or formation and classically known as La Tinta, but that several lithostratigraphic units, separated by unconformities, and of different ages, began to enlarge the classical stratigraphic scheme of the mountain range.

The first author who recognized other lithostratigraphic unit than La Tinta was Borrello (1962) who created the Punta Mogotes Formation (reinterpreted by Marchese and Di Paola 1975). Later on, other authors recognized: the Cerro Negro Formation (Iñíguez and Zalba 1974); the Cuarcitas Balcarce or the Balcarce Formation (Amos et al. 1972); the Sierras Bayas Group (Dalla Salda and Iñíguez 1979) and the amendments of Poiré, who created the Villa Mónica Formation (1987) and later, the Cerro Largo Formation (1993) within the Sierras Bayas Group; the Las Aguilas Formation (Zalba 1978) and subsequent amendments to the initial scheme: Zalba et al. (1982, 1988); the Olavarría Formation (Andreis et al. 1996) whereupon, La Tinta Formation lost identity, being divided into diverse lithostratigraphic units of different ages. Today referring to the La Tinta Formation implies going back over 153 years.

A complete compilation of the Tandilia geology and the development of the basin can be found in the classical work of Iñíguez et al. (1989). Detailed lithostratigrafic works, carried out through numerous publications in the areas of Chillar-Barker-Villa Cacique-San Manuel, are available at Zalba and Andreis (2001), field trip guide for the 12th International Clay Conference, organized by the Association International Pour l'Etude des Argilles (AIPEA), performed for the first time in Argentina, Bahia Blanca (2001).

More recent studies concerning different aspects of Tandilia (geology, stratigraphy, sedimentology, mineralogy, diagenesis, etc.) were approached by many authors. In Poiré and Spalletti (2005) and Dalla Salda et al. (2005) a very complete bibliography of papers on the Sierras de Tandilia, covering several of the aspects mentioned above can be found. These authors proposed modifications to the stratigraphic scheme of Iñíguez et al. (1989), correlating the Las Aguilas Formation with the Olavarría Formation. However, this problem is long-standing, and a detailed study is still necessary to confirm their hypothesis although some doctoral theses have been carried out in this sense but the information is still uncertain. Therefore, this paper follows the scheme of Iñíguez et al. (1989) for the stratigraphic location of the Las Aguilas Formation and the scheme of Andreis et al. (1996) for the stratigraphic location of the Olavarría Formation. In Chap. 3 we will give our reasons in favor of this proposal.

Biostratigraphic (trace fossils, stromatolites) and geochronologic (K and Rb–Sr dating) works allowed different authors to assign ages to the different lithostratigraphic units of Tandilia, and important contributions were made in this regard (cf. Borrello 1966; Antonioli 1969; Amos 1974; Rapela et al. 1974; Bonhomme and Cingolani 1978, 1980; Aceñolaza 1978; Alfaro 1981; Cingolani and Bonhomme 1982; Pöthe de Baldis et al. 1983; Poiré et al. 1984, 2003; Cingolani and Rauscher 1985; del Valle 1987; Cingolani et al. 1991; Zalba et al. 1988; Poiré and del Valle 1996), to name just a few.

The study of the crystalline basement rocks (named Complejo Buenos Aires by Di Paola and Marchese 1974), approached from different aspects by authors such as Dalla Salda (1979, 1982, 1999), Rapela et al. (1974), Teruggi and Kilmurray (1975, 1980), Varela et al. (1985, 1988), Iñiguez et al. (1989), Dalla Salda (1999) and Dalla Salda et al. (2005), among others, has been crucial in understanding the mineralogical composition of the inherited clays that integrate the different geological formations.

But this original composition of the basement rocks experimented successive hydrothermal, weathering, burial, uplift, and several stages of diagenetic processes occurred over million years ago, which dramatically modified its primary mineralogy. Already Cailtére and Iñiguez (1967) had recognized the importance of the crystalline basement as a source of supply of materials to the sediments, many of them consisting of clays. The work of Zimmermann and Spalletti (2005) and Zimmermann et al. (2005), provide information about the provenance of these clastic, detrital materials, in different lithostratigraphic units of the Tandilia System. Once deposited the detrital sediments inherited from the crystalline basement, or those generated by chemical precipitation (e.g., limestone), postdepositional processes (compaction, cementation, oxidation–reduction, erosion, rising, fracturing, introduction or expulsion of fluids, formation or transformation of minerals, etc.) occurring in the sedimentary sequences and in the basement as well, imposed a new label in the composition of the rocks. All these processes changed and enriched the mineralogy, and although many cases have been studied in detail, there are still many problems to be resolved. For example, Zalba et al. (2007) recognized phenomena of telogenesis occurred in the Middle Permian (254 million years) in Tandilia, which could be bounded in time from the radiometric K-Ar dating of K-alunite (see Chap. 3).

The lift and thrust of the Paleozoic Ventania System, located to the SW from Tandilia, were responsible for the fracturing, erosion, introduction of meteoric fluids and expulsion of trapped fluids in different areas of Tandilia through discontinuities, faults and fractures, and for the formation of an association of minerals such as kaolinite, halloysite, diaspore, alunite, and phosphates of cerium and aluminum (Ce-florencite) and of strontium and aluminum (svanvergite). The last ones are called APS minerals (aluminum-phosphate-sulfate), belong to the Alunite Supergroup, and were found in the Las Aguilas Formation, at the Cuchilla de Las Aguilas, NW of Barker town (Fig. 1.1). Nevertheless, although Dristas et al. (2003), and Martinez and Dristas (2007) also identified the presence of APS minerals, both in the basement and in the basement-sedimentary sequence unconformity, in the area of the Sierra La Juanita and at the Cuchilla de Las Aguilas (the same area studied by Zalba et al. 2007), these authors attributed them an hydrothermal origin. These important phenomena will be discussed later on the basis of field evidence and analytical data and linked to the uplift of the Paleozoic Ventania System, when the Patagonian plate and the Colorado plate collided (Ramos 1984).

According to Milani and de Wit (2008) Upper Palaeozoic stratigraphic evolution of the area was intimately linked to that of the south-western Gondwana convergent margin which was the locus for terrane accretion between Devonian and Triassic. The accretion of Patagonia (Ramos 1984, 2008; Rapalini 2005; Rapalini et al. 2010; Pángaro and Ramos 2012; Ramos and Naipauer 2014) and its possible extension into the South African Agulhas plateau (Lindeque et al. 2007) was the last colli-sional event to affect the area and resulted in the evolution of a vast foreland basin, encompassing the present day Karoo basin in South Africa, and the Claromecó (or Sauce Grande–Colorado), Carapacha and San Rafael basins in South America. Patagonia was recently interpreted by Ramos and Naipauer (2014) as a microplate detached from Antarctica and accreted to southern Gondwana between Carboniferous (Ramos 2008) and late Lower Triassic times (Pángaro and Ramos 2012). However, some authors hold different opinions about the allochthonous character of Patagonia and interpreted it as para-autochthonous (Rapalini et al. 2010) or even as autochthonous (Rapalini et al. 2013; Pankhurst et al. 2014).

Another example of important changes and mineral replacements, produced by successive post depositional processes, occurred in dolomitized limestone of the Sierra La Juanita, North East of Barker. The so called "ferruginous clays" of the Villa Mónica Formation, at the Sierra La Juanita, Estancia La Siempre Verde, Barker, largely used in the ceramic industry, and known for 40 years in the geo-logical literature with such denomination, turned out to be weathered and diage-nized dolostones, with neoformation of quartz megacrystals (of up to 20 cm long), and interstratified illite-smectite (I/S) minerals, and much later, introduction of kaolinite in fractures, at different levels (Zalba et al. 2007).

## 1.3  Geology and Stratigraphy

The Tandilia System, or Sierras Septentrionales de Buenos Aires, is located between 36° 30' and 38° South latitude and 58° and 62° West longitude. They constitute a discontinuous orographic range extending, in general, in a NW–SE direction along 300 km, from Blanca Grande to Mar del Plata (Cabo Corrientes), reaching a maximum height of 490 m above sea level, and a maximum width of 60 km in the vicinity of Tandil city (Fig. 1.1).

The central part of the Tandilia System shows a rounded and soft relief asso-ciated with outcrops of igneous-metamorphic basement rocks, while the sedimen-tary deposits spread out on a plateau, with a steep front northward (Fig. 1.2). The Tandilia sediments are composed of a set of rocks of Precambrian age in the areas

**Fig. 1.2** Rounded and soft relief, associated with outcrops of igneous-metamorphic basement rocks, while sedimentary rocks show plateau shaped outcrops

of Olavarría and Barker-San Manuel, while rocks of Lower Paleozoic age outcrop in the SE and NW sectors of the mountain range. The sediments overlie unconformably the crystalline basement rocks of the Complejo Buenos Aires, distributed in the northern part of the range, which is mainly composed of granitoids, migmatites, milonites, and ectinites; in addition, amphibolites and hypabyssal rocks, with ages ranging from 2000 to 600 million years approximately, are found in different sectors of the sierras.

The advancement of the knowledge of the sedimentary sequences of Tandilia, together with the opening of new quarries and better access to drilling data, allowed different authors to propose diverse stratigraphic schemes for different regions. The scheme adopted here is based on the proposals of Iñíguez et al. (1989), Poiré (1987, 1993) and Andreis et al. (1996), which consider six sequences or cycles of deposition, defined by relative changes in sea level. Each of these sequences is limited by regional unconformities (discontinuities). Table 1.1 corresponds to the adopted stratigraphic scheme showing, from base to top, the identified depositional sequences of four major geologic areas:

1. Sierras Bayas
2. Villa Cacique; Sierra La Juanita-Cuchilla de Las Aguilas; Sierra de La Tinta
3. San Manuel
4. Balcarce-Mar del Plata.

**Table 1.1** Stratigraphic scheme showing depositional sequences of the study areas

| Age | Area | | | | | |
| --- | --- | --- | --- | --- | --- | --- |
| | Olavarria Sierras Bayas | | Villa Cacique Loma Negra | Las Aguilas La Juanita | C° La China López Los Barrientos | Sa del Volcán Punta Mogotes |
| Early Ordovician | | | Balcarce Fm. | Balcarce Fm. | Balcarce Fm. | Balcarce Fm. |
| Neo-proterozoic | Cerro Negro Fm. | | Cerro Negro Fm. | Las Aguilas Fm. | | |
| | Sierras Bayas Group | Loma Negra Fm. | Loma Negra Fm. | | | |
| | | Olavarría Fm. | Olavarría Fm. | | | |
| | | Cerro Largo Fm. | Cerro Largo Fm. | Cerro Largo Fm. | | |
| | | Villa Mónica Fm. | Villa Mónica Fm. | La Juanita Fm. | | |
| Paleo-proterozoic | | | | | | Punta Mogotes Fm. |
| | Complejo Buenos Aires | | | | | |

Taken from Iñíguez et al. (1989), modified by Poiré (1987, 1993) and Andreis et al. (1996)

The oldest depositional sequence corresponds to the Sierras Bayas Group, comprising the following formations from base to top:

- The Villa Mónica Formation also known as the La Juanita Formation: Cuarcitas Inferiores (quartzites), dolostones (with algal mats), weathered dolostones now known as replaced by ferruginous clays (Zalba et al. 2010), siltstones and claystones.
- The Cerro Largo Formation: Cuarcitas Superiores (quartzites).
- The Olavarría Formation: (sandstones, claystones, siltstones, with Microbially Induced Sedimentary Structures (MISS).
- The Loma Negra Formation: limestones and mudstones (with MISS) and with minor clay intercalations.

In some sectors, overlying the Sierras Bayas Group it is possible to identify: the Cerro Negro Formation (limestone breccia, quartzites, claystones, siltstones, mudstones), or the Las Aguilas Formation (chert breccia, claystones, siltstones, quartzites, alternation of sandstones, and siltstones/claystones). The Sierras Bayas Group outcrops mainly in the area of Olavarría-Sierras Bayas and in the Barker area.

Finally, the sedimentary sequence of Tandilia culminates with the Balcarce Formation: quartzites with minor interbedded claystones.

For the area of Sierra La Juanita-Cuchilla de Las Aguilas-Sierra de La Tinta, the La Juanita Formation is considered a lateral variation of the Villa Mónica Formation. Instead, the Las Aguilas Formation has been considered successively a separate unit above the Loma Negra Formation (Zalba 1978); a lateral variation of the Loma Negra Formation with the same stratigraphical position as in the previous case (Iñíguez et al. 1989), and also a lateral variation of the Olavarría Formation stratigraphically located below the limestone of the Loma Negra Formation (Poiré and Spalletti 2005). All these sedimentary sequences are of Neoproterozoic age and are covered in some of these sectors, unconformably, by siliciclastic deposits (interbedded quartzites and claystones) corresponding to the Balcarce Formation, of Lower Paleozoic age (Ordovician). There is no doubt among different authors about the existence of the Balcarce Formation in the area Balcarce-Mar del Plata and in Villa Cacique. However, not all agree on the presence of this unit in the area of the Cuchilla de Las Aguilas; Sierra de La Tinta. Zalba et al. (1988) observed an angular unconformity between the deposits of the Las Aguilas Formation and the overlying quartzite deposits in the Cuchilla de Las Aguilas, ascribing the latter to the Balcarce Formation. In the Olavarría area, however, the Balcarce Formation has not been found. It either never deposited or it was completely eroded. No data are available to support any hypothesis.

The crystalline basement rocks of each area have been the main source of sediment later on carried to the Tandilia depositional basin, although postdepositional processes (erosion, uplift, fracturing, weathering, introduction of fluids, oxidation, etc.) produced transformation or neoformation of minerals. Hence local mineralogical composition and postdepositional phenomena occurred on basement rocks of different areas have produced specific associations of clay minerals and related phases.

In relation to clay deposits resulting from the "in situ" alteration of the basement rocks (residual deposits) it is important to emphasize that these are of kaolinitic-pyrophyllitic composition (Sierra La Juanita and San Manuel), or either of illitic composition (Olavarria-Sierras Bayas) when the basement is covered by sedimentary rocks of Precambrian age (Sierras Bayas Group). However, the composition of residual clay deposits is basically kaolinitic and illitic, with subordinated superimposed interstratified illite-smectite, when the weathered basement rocks are covered by Paleozoic sediments (Balcarce Formation). These facts were already recognized by Iñíguez et al. (1989). Definitely, the diagenetic changes that are superimposed differ in each case.

# References

Aceñolaza FG (1978) El Paleozoico inferior de Argentina según sus trazas fósiles. Ameghiniana 15:15–64

Alfaro MB (1981) Estudio geológico de la zona comprendida por las Hojas La Numancia, Licenciado Matienzo y Estancia San Antonio, en las Sierras Septentrionales de la provincia de Buenos Aires. 5° Reunión Científica Informativa, Resúmenes: 9, Comisión de Investigaciones Científicas (C.I.C.) de la provincia de Buenos Aires, La Plata

Amos AJ (1974) Los estromatolitos del Precámbrico sedimentario de la Formación La Tinta, Provincia de Buenos Aires. LEMIT, La Plata 2(269):151–155

Amos A, Quartino B, Zardini R (1972) Grupo "La Tinta" (Provincia de Buenos Aires, Argentina) Paleozoico o Precámbrico? In: Proc 25° Congresso Brasileiro de Geología, Sao Paulo, Brasil pp 211–221

Andreis RR, Zalba PE, Iñíguez AM, Morosi M (1996) Estratigrafía y evolución paleoambiental de la sucesión superior de la Formación Cerro Largo, Sierras Bayas (Buenos Aires, Argentina). In: Proc 6° Reunión Argentina de Sedimentología, pp 293–298

Antonioli JA (1969) Formación La Tinta. Notas de la Comisión de Investigaciones Científicas, Provincia de Buenos Aires, 6(5):1–32

Bonhomme MG, Cingolani CA (1978) First isotopic dating of upper precambrian sediments in the province of Buenos Aires. In: Short papers of the Fourth International Conference, Geochronology, Cosmochronology, Isotope Geology, Geological Survey, USA, Open-File Report 78-701, pp 45–46

Bonhomme MG, Cingolani CA (1980) Mineralogía y geocronología Rb/Sr y K/Ar de fracciones finas de la Formación La Tinta, Provincia de Buenos Aires. Rev Asoc Geol Argentina 35 (4):519–538

Borrello AV (1962) Formación Punta Mogotes (Eopaleozoico—Provincia de Buenos Aires). Notas Comisión de Investigación Científica Provincia de Buenos Aires 1(1):1–9

Borrello AV (1966) Trazas, restos tubiformes y cuerpos fósiles problemáticos de la Formación La Tinta, Sierras Septentrionales de la Provincia de Buenos Aires. Comisión de Investigaciones Científicas de la Provincia de Buenos Aires. Paleontografía Bonaerense 5:1–42

Caillére S, Iñíguez AM (1967) Etude minerlogique de "La Tinta Formation" argilleuse de la Province de Buenos Aires, Republique Argentine. Bull Soc Franc Min Cristall, Paris 90:245–251

Cingolani CA, Bonhomme MG (1982) Geochronology of La Tinta upper proterozoic sedimentary rocks, Argentina. Precambrian Res 18(1–2):119–132

Cingolani C, Rauscher R (1985) Datos geocronológicos en las sedimentitas del Grupo La Tinta de Villa Cacique, Partido de Juárez, Provincia de Buenos Aires. 1° Jornadas Geológicas Bonaerenses, Tandil, 1985. Publicaciones Comisión de Investigaciones Científicas, La Plata, p 128

Cingolani CA, Rauscher R, Bonhomme M (1991) Grupo La Tinta (Precámbrico y Paleozoico Inferior), provincia de Buenos Aires, Argentina: Nuevos datos geocronológicos y micropaleontológicos en las sedimentitas de Villa Cacique, partido de Juárez. Rev Téc YPFB, Bolivia 12(2):177–191

Dalla Salda LH (1979) Nama and La Tinta groups, a common Southern Africa-Argentina basin? In: Bull Chamber of Mines Precambrian Research Unit, University of Cape Town, 16th Annual Report, pp 113–128

Dalla Salda LH (1982) Nama-La Tinta y el inicio de Gondwana. Acta Geológica Lilloana 16 (1):23–28

Dalla Salda LH (1999) Cratón del Río de la Plata, El Basamento granítico-metamórfico de Tandilia y Martín Gracia. In: Caminos R (ed) Geología Regional Argentina. Anales del Instituto de Geología y Recursos Minerales (SEGEMAR) 29, pp 97–100

Dalla Salda LH, Iñíguez AM (1979) La Tinta, Precámbrico y Paleozoico de Buenos Aires. In: Proc 7° Congreso Geológico Argentino, Neuquén, vol 1, pp 539–550

Dalla Salda L, Guichon M, Rapela C (1972) Hallazgo de una brecha de talud en el techo de las calizas de Barker, Provincia de Buenos Aires, República Argentina. Rev Asoc Arg Min, Petrol Sedim 3(3–4):133

Dalla Salda LH, de Barrio RE, Echeveste HJ, Fernández RR (2005) El basamento de las Sierras de Tandilia. In: Geología y Recursos minerales de la Provincia de Buenos Aires, de Barrio RE, Etcheverry RQ, Caballé MF, Llambías EJ (eds) Proc 16° Congreso Geológico Argentino, La Plata, vol 3, pp 31–50

del Valle A (1987) Nuevas trazas fósiles en la Formación Balcarce, Paleozoico inferior de las Sierras Septentrionales. Su significado cronológico y ambiental. Rev Museo de La Plata, nueva serie, Sección Paleontología 9:19–41

Di Paola E, Marchese HG (1974) Relación entre la tecto-sedimentación litología y mineralogía de arcillas del Complejo Buenos Aires y la Formación la Tinta (prov. de Buenos Aires). Rev Asoc Arg Min, Petrol Sedim, Buenos Aires, 5(3–4):45–58

Dristas JA, Frisicale MC, Martínez JC (2003) High-REE APS minerals associated with advanced argillic alteration in the Cerrito de la Cruz deposit, Barker, Buenos Aires province, Argentina. Göttinger Arbeiten für Geologie und Paläontologie, Sb (Festschrift Behr), pp 1–6

Heusser J, Claraz G (1863) Ensayo de un conocimiento geognóstico físico de la Provincia de Buenos Aires 1. La cordillera entre Cabo Corrientes y Tapalqué, Buenos Aires

Holmberg E (1972) Tandilia. In: Leanza A (ed) Geología Regional Argentina. Academia Nacional de Ciencias, Córdoba, pp 365–393

Iñiguez MA, Zalba PE (1974) Nuevo nivel de arcilitas en la zona de Cerro Negro, partido de Olavarría, provincia de Buenos Aires. LEMIT Serie 2(264):95–100

Iñiguez AM, Del Valle A, Poiré D, Spalletti L, Zalba P (1989) Cuenca Precámbrica-Paleozoico inferior de Tandilia, Provincia de Buenos Aires. In: Chebli G, Spalletti LA (eds) Cuencas sedimentarias argentinas. Instituto Superior de Correlación Geológica, Universidad Nacional de Tucumán, Serie Correlación Geológica, 6:245–263

Lindeque A, Ryberg T, Stankiewicz J, Weber M, de Wit MJ (2007) Deep crustal seismic reflection experiment across the southern Karoo Basin, South Africa. S Afr J Geol 110:419–438

Marchese HG, Di Paola E (1975) Reinterpretación estratigráfica de la perforación de Punta Mogotes I, Provincia de Buenos Aires. Rev Asoc Geol Argentina 30(1):44–52

Martinez JC, Dristas J (2007) Paleoactividad hidrotermal en la discordancia entre el Complejo Buenos Aires y la Formación La Tinta en el área de Barker, Tandilia. Rev Asoc Geol Argentina, Buenos Aires 62(3):375–386

Milani EJ, de Wit MJ (2008) Correlations between the classic Parana' and CapeKaroo sequences of South America and southern Africa and their basin infills flanking the Gondwanides: du Toit revisited. In: Pankhurst, RJ, Trouw, RAJ, Brito Neves, BB, De Wit, MJ (eds), West Gondwana: Pre-Cenozoic Correlations Across the South Atlantic Region. Geological Society, Special Publications, London, vol 294, pp 319–342

Nágera JJ (1940) Tandilia. In: Historia Física de la Provincia de Buenos Aires. Revista Humanidades I, pp 1–272

Pángaro F, Ramos VA (2012) Paleozoic crustal blocks of onshore and offshore central Argentina: new pieces of the southwestern Gondwana collage and their role in the accretion of Patagonia and the evolution of Mesozoic south Atlantic sedimentary basins. Mar Pet Geol 37(1):162–183

Pankhurst RJ, Rapela CW, López de Luchi MG, Rapalini AE, Fanning CM, Galindo C (2014) The Gondwana connections of northern Patagonia. J Geol Soc 171:313–328

Poiré DG (1987) Mineralogía y sedimentología de la Formación Sierras Bayas en el núcleo septentrional de las sierras homónimas, Olavarría provincia de Buenos Aires. Tesis Doctoral 494 (inédito). Facultad de Ciencias Naturales y Museo, Universidad Nacional de La Plata, p 271

Poiré DG (1993) Estratigrafía del Precámbrico sedimentario de Olavarría Sierras Bayas, provincia de Buenos Aires, Argentina. In: Proc 13° Congreso Geológico Argentino y 3° Congreso de Exploración de Hidrocarburos, Mendoza, vol 2, pp 1–11

Poiré DG, del Valle A (1996) Trazas fósiles en barras submareales de la Formación Balcarce (Ordovícico), Cabo Corrientes, Mar del Plata, Argentina. Asociación Paleontológica Argentina, Publicación Especial, vol 4, pp 89–102

Poiré DG, Spalletti LA (2005) La cubierta sedimentaria Precámbrica-Paleozoica inferior del Sistema de Tandilia. In: de Barrio RE, Etcheverry RO, Caballé MF, Llambías E (eds) Geología y Recursos Minerales de la Provincia de Buenos Aires. Proceedings 16° Congreso Geológico Argentino, Relatorio 4, La Plata, pp 51–68

Poiré DG, del Valle A, Regalía GM (1984) Trazas fósiles en cuarcitas de la Formación Sierras Bayas (Precámbrico) y su comparación con las de la Formación Balcarce (Cambro-Ordovícico), Sierras Septentrionales de la provincia de Buenos Aires. In: Proc 9° Congreso Geológico Argentino, vol 4, pp 249–266

Poiré DG, Spalletti LA, del Valle A (2003) The Cambrian-Ordovician siliciclastic platform of the Balcarce Formation (Tandilia System, Argentina): Facies, trace fossils, palaeoenvironments and sequence stratigraphy. Geologica Acta 1:41–60

Pöthe de Baldis ED, Baldis B, Cuomo J (1983) Los fósiles precámbricos de la Formación Sierras Bayas (Olavarría) y su importancia intercontinental. Rev Asoc Geol Argentina 38(1):73–83

Ramos VA (1984) Patagonia. Un continente a la deriva? In: Proc 9° Congreso Geológico Argentino, Bariloche, vol 2, pp 311–325

Ramos VA (2008) Patagonia: a Paleozoic continent adrift? J S Am Earth Sci 26:235–251

Ramos VA, Naipauer M (2014) Patagonia: where does it come from? J Iber Geol 40(2):367–380

Rapalini AE (2005) The accretionary history of southern South America from the latest Proterozoic to the Late Palaeozoic: some palaeomagnetic constraints. In: Vaughan AEM, Leat PT, Pankhurst RJ (eds) Terrane processes at the Margins of Gondwana, vol 246. Geological Society, Special Publications, London, pp 305–328

Rapalini AE, López de Luchi M, Martínez Dopico C, Lince Klinger F, Giménez M, Martínez P (2010) Did Patagonia collide with Gondwana in the late Paleozoic? Some insights from a multidisciplinary study of magmatic units of the North Patagonian Massif. Geol Acta 8:349–371

Rapalini AE, López de Luchi M, Tohver E, Cawood PA (2013) The South American ancestry of the North Patagonian Massif: geochronological evidence for an autochthonous origin? Terra Nova 25:337–342

Rapela C, Dalla Salda LH, Cingolani C (1974) Un filón básico ordovícico en la Formación La Tinta, Sierra de los Barrientos Provincia de Buenos Aires. Rev Asoc Geol Argentina 29 (3):319–331

Teruggi ME, Kilmurray JO (1975) Tandilia. Relatorio Geología de la Provincia de Buenos Aires. In: Proc 6° Congreso Geológico Argentino, pp 55–78

Teruggi M, Kilmurray J (1980) Sierras Septentrionales de la provincia de Buenos Aires. In: Turner JCM (ed) Proc 2° Simposio Geología Regional Argentina. Academia Nacional de Ciencias de Córdoba, Córdoba, Argentina II, pp 919–956

Varela R, Dalla Salda L, Cingolani C (1985) La edad Rb-Sr del Granito de Vela, Tandil. In: Proc 1° Jornadas Geológicas Bonaerenses (Tandil) Comisión Investigaciones Científicas, provincia de Buenos Aires, La Plata, Argentina, pp 881–891

Varela R, Cingolani CA, Dalla Salda LH (1988) Geocronología rubidio-estroncio en granitoides del basamento de Tandil, provincia de Buenos Aires, Argentina. In: Proc 2° Jornadas Geológicas Bonaerenses (Bahía Blanca). Comisión Investigaciones Científicas, provincia de Buenos Aires, La Plata, Argentina, pp 291–305

Zalba PE (1978) Estudio geológico-mineralógico de los yacimientos de arcillas de la zona de Barker, partido de Juárez, Provincia de Buenos Aires y su importancia económica. Tesis Doctoral 362, (inédito). Facultad de Ciencias Naturales y Museo, Universidad Nacional de La Plata, p 75

Zalba PE, Andreis RR (2001) Stratigraphy, sedimentology and mineralogy of Neoproterozoic clay deposits, Sierras de Tandilia, Province of Buenos Aires, Argentina. Economical Importance. 12th International Clay Conference, Pre-Simposium Field Trip, Bahía Blanca, p 80

Zalba PE, Andreis RR, Lorenzo F (1982) Consideraciones estratigráficas y paleoambientales de la secuencia basal eopaleozoica en la Cuchilla de Las Aguilas, Barker, Argentina. In: Proc 5° Congreso Latinoamericano de Geología Argentina, Buenos Aires 2, pp 389–409

Zalba PE, Andreis RR, Iñíguez AM (1988) Formación Las Aguilas, Sierras Septentrionales de Buenos Aires, nueva propuesta estratigráfica. Rev Asoc Geol Argentina, Buenos Aires 43 (2):198–209

Zalba PE, Manassero M, Laverret EM, Beaufort D, Meunier A, Morosi M, Segovia L (2007) Middle Permian telodiagenetic processes in Neoproterozoic sequences, Tandilia System, Argentina. J Sed Res 77:525–538

Zalba PE, Manassero M, Morosi ME, Conconi MS (2010) Preservation of biogenerated mixed facies: a case study from the Neoproterozoic Villa Mónica Formation, Sierra La Juanita, Tandilia, Argentina. J Appl Sci 10(5):363–379

Zimmermann U, Spalletti LA (2005) The provenance of the lower Palaeozoic Balcarce formation (Tandilia System, Buenos Aires Province, Argentina). In: Proc 16 Congreso Geológico Argentino, La Plata, vol 3, pp 203–210

Zimmermann U, Poiré DG, Gómez Peral L (2005) Provenance studies on Neoproterozoic successions of the Tandilia System (Buenos Aires Province, Argentina): preliminary data. In: Proc 16° Congreso Geológico Argentino, vol 4, pp 561–568

# Chapter 2
# Azul County

**Abstract** This chapter is related to the geology, stratigraphy, petrology, mineralogy, and industrial applications of residual clay deposits of Azul County, Chillar Sector. Almost all of the outcrop rocks in the area of Azul correspond to the crystalline basement rocks (Complejo Buenos Aires), 2200–1800 Ma, belonging to the Trans Amazonic cycle (Teruggi and Kilmurray 1975, 1980; Dalla Salda et al. 1988, 1992). The mountain area is characterized by the presence of middle and high metamorphic grade (gneisses and migmatites) and of granitoid rocks, sometimes with variable degree of milonitization, produced by E–W-oriented shear zones. The residual clay deposits are weathering products of the crystalline basement rocks and are covered by siliciclastic sediments of the Balcarce Formation (oligomictic orthoconglomerates and quartzites). In the area of Azul, the clay deposits are located at the Estancia Santa María and La Verónica, in the vicinity of the national route N° 3; 3 and 5 km to the SSE of the town of Chillar, respectively. Chillar clays are mineralogically classified as kaolinitic, and technologically, as refractory of high and medium quality, in coincidence with the high percentages of $Al_3O_2$ observed in the chemical analyses and the abundant content of kaolinite with subordinate dickite as predominant clay minerals in the saprock and in the saprolite as well.

**Keywords** Azul County · Chillar Sector · Geology · Stratigraphy · Mineralogy · Residual clay deposits · Technological properties

The County of Azul bounds to the North with Tapalqué and Las Flores counties, to the South with Benito Juárez County, to the East with Rauch and Tandil counties, and to the West with the Olavarría County. Almost all of the outcrop rocks in the area of Azul correspond to the crystalline basement rocks (Complejo Buenos Aires), 2200–1800 Ma, belonging to the Trans Amazonic cycle (Teruggi and Kilmurray 1975, 1980; Dalla Salda et al. 1988, 1992). The mountain area is characterized by the presence of middle and high metamorphic grade (gneisses and migmatites) and of granitoid rocks, sometimes with variable degree of milonitization, produced by E–W-oriented shear zones.

© Springer International Publishing Switzerland 2016
P.E. Zalba et al., *Gondwana Industrial Clays*,
Springer Earth System Sciences, DOI 10.1007/978-3-319-39457-2_2

15

In addition, conspicuous thick bands of cataclasites, mainly of granitic composition, are observed. In this set of rocks that make up the local metamorphic basement, there are regular size diabase dykes, often anphibolitized, presenting a predominantly NW direction (Dalla Salda 1981; Kilmurray et al. 1985; Etcheveste et al. 1997). Moreover, the Paleozoic sedimentary sequence (The Balcarce Formation, see Table 1.1) is restricted to the sector of Chillar (60 km S from Azul and 75 km to the west of Tandil. In this area the Balcarce Formation consists of variable thickness (between 8 and 20 m) of conglomerates and sandstones which overlie a weathered (saprolitized) crystalline basement rock. Oligomictic orthoconglomerates consist of very thick to thick tabular or lenticular strata, of light colors (gray, whitish) and sub-rounded clasts of chert composition (microcrystalline silica = chalcedony). The entire sequence, which is approximately 8 m thick in the area, culminates with quartzites, with abundant cross-bedding structures. The structure sinks regionally, with low-angle, towards the SE.

## 2.1   Chillar Sector

### 2.1.1   Residual Deposits: Characteristics, Mineralogical, and Chemical Composition

In the area of Azul, the clay deposits are located at the Estancia Santa María and La Verónica, in the vicinity of the national route Nº 3; 3 and 5 km to the SSE of the town of Chillar, respectively. These deposits are weathering products of the crystalline basement rocks (residual deposits) and are covered by siliciclastic sediments of the Balcarce Formation (oligomictic orthoconglomerates and quartzites).

La Verónica is the main site of exploitation where the extracted clay level represents a saprolitized zone (Fig. 2.1a) of the underlying crystalline basement rocks with a thickness varying between 1 and 4 m. Figure 2.1b represents a stratigraphic section in the La Verónica quarry, where the different areas of weathering of the basement rocks can be seen and, unconformably, over an erosive discordance, the sedimentary sequence of the Balcarce Formation lies.

According to petrographic analysis the base of the section corresponds to the weathering zone known as saprock since the excavation of the quarry does not reach the unaltered basement (bedrock). The saprock is recognized because, although the original mineral species remain as such, the rock already has a recognizable degree of alteration. The mineral species identified at this level are: plagioclase (oligoclase: An 17–22 %), some of them twinned and without alteration, and the most abundant, with no twins, have altered to kaolinite-dickite and, to a lesser extent, to illite-smectite, which appears as whitish dots (up to yellowish)

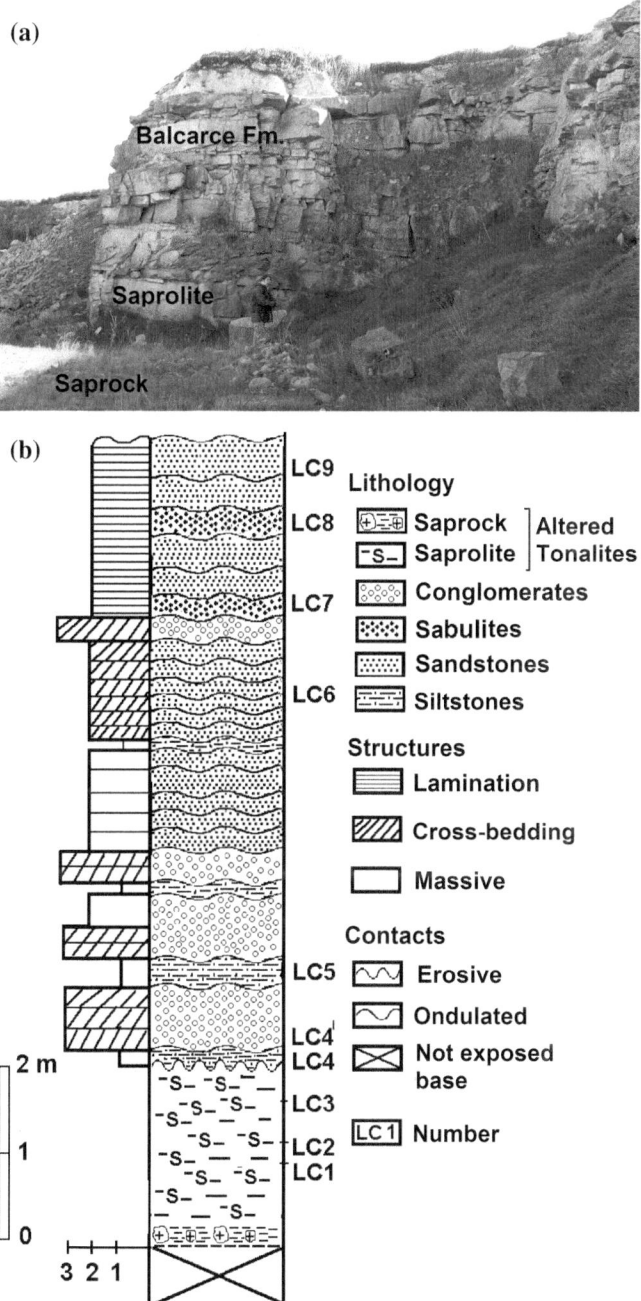

**Fig. 2.1 a** La Verónica quarry, Chillar. The clays mined correspond to the saprolite (altered basement). **b** Stratigraphic section carried out in the La Verónica quarry (Chillar), showing different alteration basement zones and the sedimentary sequence (Balcarce Formation). LC1/9: samples

**Fig. 2.2** Photomicrographs of La Verónica quarry, Chillar. Saprock: **a** Plagioclase altered to kaolinite-dickite (K-D) and scarce fresh microcline (Mc), (40x magnification). **b** Monocrystalline quartz with undulate extinction (Qz), Biotite (Bt) in the form of curved books of high birefringence (40x magnification). **c** Iron exhaust through open cleavages in biotite (Bt), (200x magnification). **d** Poorly defined kaolinite books (K), ghostly, (200x magnification). Saprolite: **e** Migration of hematite (He) into cracks and fissures, (200x magnification). Sedimentary level: **f** Monocrystalline quartz (Qm) and fresh hornblende (Hb), (100x magnification). Photomicrographs taken under cross polarized light (XPL)

within the plagioclase (Fig. 2.2a). Also the presence of scarce, fresh microcline was observed. Monocrystalline quartz (Fig. 2.2b) as well as abundant curved mica (biotite, Fig. 2.2c) of high birefringence, and which begins to alter to kaolinite, distinguished by its grayish colors are observed. The cleavages are opened and constitute the escape route of the iron contained in the mica structure. Zircon was identified as an accessory mineral.

Going upwards in the section, the level known as saprolite is reached. Here the original minerals (plagioclase, biotite) do not exist anymore. The alteration to clay minerals is complete. Kaolinite books, mostly, lose definition and become "ghostly" (Fig. 2.2d). The only relic of biotite is iron oxide (hematite), in some cases forming a laminar texture as vestige of the original mica texture, or either crystallizing into prismatic shapes (Fig. 2.2e). Hematite migrates to the cracks and fissures as well as the clays, which, in addition, surround quartz crystals (Fig. 2.2e). The altered basement rock (saprolite) is compact, with conchoidal fracture and reddish-gray to gray-green colored, occasionally reddish to the base due to the presence of iron oxides and hydroxides. The greenish gray coloration is due to the clay minerals. This is the exploited level of the quarry.

On the other hand, the clays of the first sedimentary levels, above the con-glomerate of the Balcarce Formation (not exploited) show a completely different texture, with unaltered muscovites, monocrystalline quartz, fresh hornblende (Fig. 2.2f), and clays. The mineralogical composition of the clays in the La Verónica, analyzed by X-ray diffraction for this work (Table 2.1), is consistent with the interpretation of the petrographic analysis. It is essentially kaolinitic, in some sectors with high degree of purity, accompanied by scarce interstratified clay minerals: illite-smectite (IS), and very little smectite. The most common impurities are iron oxides and quartz, the latter found in varying proportions (12–56 %) in angular crystals, rarely exceeding 1 mm. Clays show a high degree of compaction. The presence of anatase denotes a high degree of weathering.

As for the Estancia Santa Maria (Table 2.2, samples SM) the base rock is of the same type and the same thickness as the one described at the La Verónica, although with some variations. Figure 2.3 shows a front of exploitation. The rock base corresponds to a level of weathered basement (saprock). Petrographic studies car-ried out for this study shows that black micas (biotites) are abundant while white micas (muscovites) are scarce.

Biotites are partially deferrized, with yellowish interference colors and varying degrees of alteration (Fig. 2.4a). Some crystals of biotite are flexured and oriented

**Table 2.1** Mineralogical composition by X-ray diffraction, temperature and Pyrometric Cone Equivalent (PCE) of the studied samples in the La Verónica quarry. Samples LC 1-4 correspond to altered levels of basement. LC 5 is the first sedimentary clay level above the conglomerate of the Balcarce Formation

| Sample | Quartz (%) | Kaolinite (%) | I/S (%) | Smectite (%) | Anatase (%) | Temperature (°C) | PCE |
|--------|-----------|---------------|---------|--------------|-------------|------------------|------|
| LC1 | 14 | 77 | 6 | 3 | – | 1680 | 31 |
| LC2 | 12 | 80 | 5 | 3 | – | 1765 | 34 |
| LC3 | 17 | 75 | 5 | 3 | Traces | 1700 | 31.5 |
| LC4 | 24 | 70 | 4 | 2 | Traces | 1665 | 30 |
| LC5 | 21 | 73 | 3 | 2 | 1 | 1680 | 31 |

**Table 2.2** Mineralogical composition by X-ray diffraction, temperature and Pyrometric Cone Equivalent (PCE) of the studied samples in the Santa Maria quarry. Samples SM 0-2 correspond to residual deposits and SM3 corresponds to the sedimentary level of the Balcarce Formation

| Sample | Quartz (%) | Kaolinite (%) | I/S (%) | Smectite (%) | Anatase (%) | Temperature (°C) | PCE |
|--------|-----------|---------------|---------|--------------|-------------|------------------|-----|
| SM0 | 56 | – | 43 | 1 | – | 1621 | 26 |
| SM1 | 39 | 56 | 4 | 1 | – | 1680 | 31 |
| SM2 | 44 | 50 | 5 | 1 | – | 1680 | 31 |
| SM3 | 38 | 54 | 5 | 1 | 2 | 1640 | 27 |

**Fig. 2.3** Santa María quarry, Chillar. The clay level mined corresponds to the saprolite. The saprock forms the base of the quarry

around abundant monocrystalline quartz crystals. Ferric oxide (product of the deferrization of biotite) is located in its open cleavages and occupying intracrystalline and intercrystalline fractures and fissures. The plagioclase in this level of the saprock is very abundant, most of it is fresh and an incipient alteration to kaolinite-dickite is observed. The accessory minerals are crystals of zircon, in some cases, zoned. Following up on the profile the saprolite zone is reached where abundant plagioclase, with varying degrees of alteration to clay minerals such as kaolinite; dickite; illite-smectite (I/S) in the form of grayish to yellowish aggregates, mostly along the cleavages (Fig. 2.4b) are identified. Minor biotites are also deferrized and altered to clay minerals. Iron oxides are arranged on the lines of

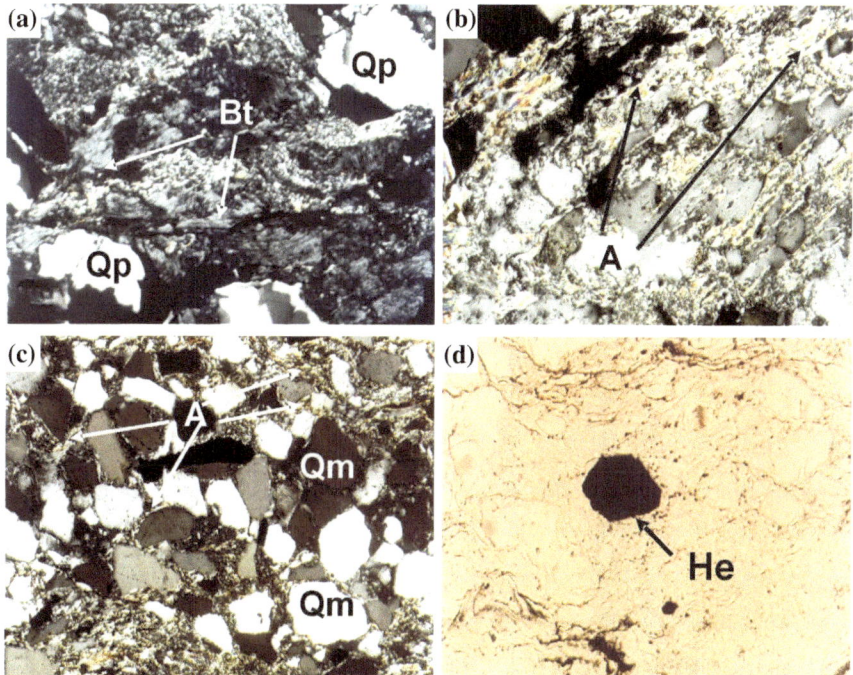

**Fig. 2.4** Photomicrographs of Santa Maria quarry, Chillar. **a** Saprock: biotites partly deferrized and flexured (Bt); polycrystalline quartz (Qp), XPL, 200x magnification. **b** Saprolite: clay minerals (A) forming aggregates along the cleavage (K-D), XPL, 200x magnification. Sedimentary level: **c** laminates levels with alternating abundant sandy or abundant clay. Monocrystalline quartz crystals (Qm) surrounded by clay (A), XPL, 200x magnification. **d** Hematite hexagon (I) cross section. Photomicrograph taken under plane polarized light (PPL), 100x magnification

cleavage of biotite or forming irregular aggregates in pore spaces. Zircon is an accessory component. Clays migrate to fissures and surround abundantly various mineral species.

Continuing upward in the stratigraphic section, and above the conglomerate of the Balcarce Formation, sandy siltstone levels are recognized. A drastic change in the texture of the rock is detected. The deposits are laminated, with alternating abundant sandy or clayish levels. Monocrystalline quartz is very abundant and detrital clay coatings surround the clasts (Fig. 2.4c). Iron oxide crystallizes as irregular or prismatic hematite, and also it is observed in basal, hexagonal sections (Fig. 2.4d). Zircon crystals increase and are of rounded shapes. The presence of anatase in the sedimentary levels is a product of its remobilization from the crystalline rocks, formed in the upper part of the saprolite. These levels are not exploited.

The mineralogical composition by X-ray diffraction (XRD) of samples from the saprolite, studied in the Santa María quarry, is available in Table 2.2. Lateral variations in the mineralogical composition are appreciated taking into account the two mentioned deposits, especially at lower levels, corresponding to the saprock. Quartz is significantly more abundant at the Santa María, probably due to local variations of the original rock. The variation in the amount of quartz in LC5 and SM3 (sedimentary levels) is due to erosion and remobilization of the transported materials. On the basis of petrographic analyses, the basement rocks (bedrock) of both stratigraphic sections would correspond to biotitic tonalites.

Chemical analyses on samples from La Verónica and Santa María can be consulted in Table 2.3. Variable content of $Al_2O_3$ (15–31.6 %) is due to the destruction of plagioclase and muscovite and to the concentration of $Al_2O_3$ in the clay minerals of upper levels of the altered basement rocks. $SiO_2$ contents correspond to the addition of silica to the new formed clay minerals and the crystallization of quartz in cracks, except in SM3 and LC5 (sedimentary levels) which reflect erosion and transport processes.

The low $Fe_2O_3$ content (0.4–1.24 %) is explained because iron is the first element which solubilizes in a weathering profile. The low Na, Mg, and Ca contents and their variations, correspond, also, to the loss of these elements in a weathering profile, except in samples LC5 and SM3 (sedimentary levels) where their increase is due to the presence of unaltered muscovite transported to the basin. In terms of the variation in K content, this is consistent with the destruction of the muscovites as weathering progresses in the basement rocks. Its increase in LC5 and SM3 (sedimentary levels) is coherent with the presence of unaltered muscovite transported to

**Table 2.3** Chemical composition of samples of La Verónica (LV) and Estancia Santa Maria (SM) quarries

| Sample | LC1 | LC2 | LC3 | LC4 | LC5 | SMO | SM1 | SM2 | SM3 |
|---|---|---|---|---|---|---|---|---|---|
| Weigth % | | | | | | | | | |
| $SiO_2$ | 51.1 | 53.8 | 56.2 | 53.3 | 54.2 | 76.1 | 72.6 | 71.2 | 61.3 |
| $Al_2O_3$ | 31.9 | 30.0 | 28.9 | 31.6 | 30.0 | 15.1 | 18.85 | 19.55 | 23.0 |
| $Fe_2O_3$ | 1.24 | 1.00 | 1.07 | 1.12 | 0.70 | 0.43 | 0.44 | 0.43 | 0.69 |
| CaO | 0.21 | 0.30 | 0.16 | 0.17 | 0.41 | 0.21 | 0.09 | 0.30 | 0.61 |
| MgO | 0.20 | 0.24 | 0.14 | 0.15 | 0.26 | 0.17 | 0.14 | 0.14 | 0.44 |
| $Na_2O$ | 0.02 | 0.01 | 0.01 | 0.01 | 0.06 | 0.02 | <0.01 | 0.01 | 0.03 |
| $K_2O$ | 0.83 | 0.80 | 0.41 | 0.38 | 0.63 | 3.89 | 0.85 | 0.67 | 1.34 |
| $Cr_2O_3$ | 0.01 | 0.02 | 0.03 | 0.01 | 0.02 | <0.01 | <0.01 | 0.01 | 0.03 |
| $TiO_2$ | 1.79 | 1.34 | 1.19 | 1.08 | 1.25 | 0.37 | 0.39 | 0.36 | 2.26 |
| MnO | 0.01 | 0.01 | <0.01 | 0.01 | 0.01 | 0.05 | <0.01 | 0.01 | <0.01 |
| $P_2O_5$ | <0.01 | <0.01 | 0.07 | < 0.01 | 0.01 | 0.06 | <0.01 | 0.02 | 0.09 |
| SrO | 0.01 | <0.01 | 0.01 | <0.01 | 0.01 | 0.01 | <0.01 | <0.01 | 0.01 |
| BaO | 0.02 | 0.02 | 0.01 | 0.02 | 0.02 | 0.03 | 0.01 | 0.05 | 0.05 |
| $H_2O-$ | 0.85 | 1.16 | 0.70 | 1.05 | 0.98 | 0.70 | 0.39 | 0.44 | 1.23 |
| $H_2O+$ | 11.75 | 11.35 | 11.05 | 12.5 | 10.9 | 2.40 | 6.55 | 6.97 | 7.82 |
| LOI | 12.45 | 12.25 | 11.35 | 12.6 | 12.4 | 3.24 | 6.88 | 7.32 | 9.47 |
| Total | 99.8 | 99.8 | 99.6 | 100.5 | 100 | 99.7 | 100 | 100 | 99.3 |

the basin. The $TiO_2$ increase in sedimentary levels (LC5 and SM3) is due to its remobilization, originally contained in the muscovite, and to the formation of anatase, and, also, to the transport of fresh muscovite to the basin.

## 2.1.2 Scanning Electron Microscopy of Clays of La Verónica

Texture and mineralogical composition of the saprock from the La Verónica is exemplified by SEM. Incipient (very rarely observed) pseudo hexagonal kaolinite crystals formed from feldspar (plagioclase) are shown. The crystals are grouped into "accordions" or "books", typical of neoformed kaolinite. Also, in the right half corner of the picture, a crystal of goethite (G), with serrated edges (Fig. 2.5a) is observed. Figure 2.5b also belongs to the saprock and shows very thin goethite

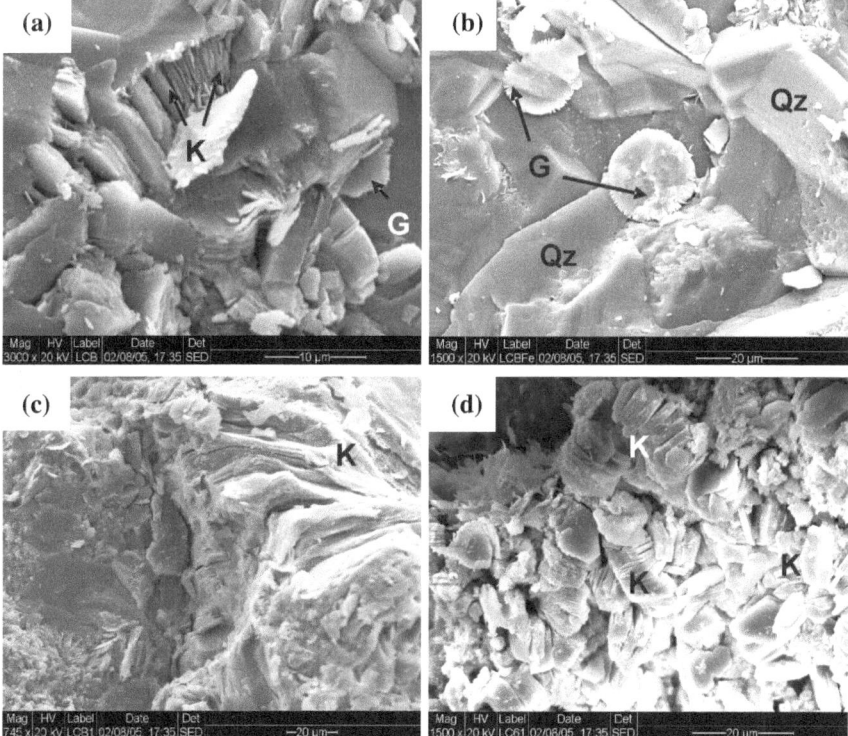

**Fig. 2.5** Scanning electron micrographs. La Verónica quarry. Saprock: **a** Kaolinite crystals growing from a plagioclase. **b** Goethite crystals growing over quartz crystals. Saprolite: **c** Partly deferrized micas through open cleavages and transformed into kaolinite. Sedimentary level: **d** Accordions of pseudo hexagonal diagenetic kaolinite. The bar indicates the scale of the microphotographs

crystals growing over quartz crystals, indicating the paragenetic sequence. In the saprolite, packages of ex-micas, partly deferrized, through open cleavages, and transformed into kaolinite can be seen in Fig. 2.5c. On the other hand, Fig. 2.5d exhibits the diagenetic growth of regular, pseudohexagonal kaolinite "accordions", as well as of illite-smectite (upper left sector of the image) of filamentous forms, in the pores of sandstone (quartzites), in the sedimentary deposits of the Balcarce Formation, overlying unconformably the saprolite.

## 2.1.3   Technological Properties of Residual Clays of La Verónica and Santa Maria

On the basis of mineralogical and physico-chemical analyses carried out by the authors for this work, and those taken from the geological literature (Schalamuk et al. 1992), it can be concluded that the clays of the saprolite in the area of Chillar have scarce variations. The plasticity index is between 3 and 5, the bulk density is between 2.05 and 2.1, the apparent porosity varies between 18.3 and 23.2 % and the Pyrometric Cone Equivalent (PCE) is between 26 and 34. The refractoriness of a material is inversely proportional to the alkalis content, and values of refractoriness are presented in Table 2.4. The PCE analysis for refractory materials indicates the temperature at which the material becomes soft by the action of heat. The essay was carried out according to the method detailed in the standard IRAM 12,507. The color of the piece after burning is ivory white, which will result in a clear ceramic as a final product.

Chillar clays are classified as refractory with high (PCE 31–34) and medium (PCE 26–27) quality, in coincidence with the high percentages of $Al_3O_2$ observed in the chemical analyses and the abundant content of kaolinite with subordinate dickite as predominant clay minerals in the saprock and in the saprolite as well.

**Table 2.4** Pyrometric Cone Equivalent (PCE) and temperature of the analyzed samples of La Verónica and Santa Maria quarries

| Sample | Temperature (°C) | P.C.E. |
|--------|------------------|--------|
| LC1    | 1680             | 31     |
| LC2    | 1765             | 34     |
| LC3    | 1700             | 31.5   |
| LC4    | 1665             | 30     |
| LC5    | 1680             | 31     |
| SMO    | 1621             | 26     |
| SM1    | 1680             | 31     |
| SM2    | 1680             | 31     |
| SM3    | 1640             | 27     |

# References

Dalla Salda LH (1981) Tandilia, un ejemplo de tectónica de transcurrencia en basamento. Rev Asoc Geol Argentina 36(2):204–207

Dalla Salda LH, Bossi J, Cingolani CA (1988) The Rio de la Plata cratonic region of southwestern Gondwana. Episodes 11(4):263–269

Dalla Salda LH, Franzese JR, Posadas VG (1992) The 1800 Ma mylonite-anatectic granitoid association in Tandilia, Argentina. Basement Tectonics 7:161–174

Etcheveste H, Fernandez R, Ribot A, Teixeira W (1997) Nuevos fechados radimétricos Rb/Sr para diques del Sistema de Tandilia. Jornadas de comunicaciones Científicas, Facultad de Ciencias Naturales y Museo, La Plata, pp 217

Kilmurray JO, Leguizamón MA, Ribot A (1985) Los diques de diabasa del noroeste de las Sierras de Azul, Provincia de Buenos Aires. In: Proc 1° Jornadas Geológicas Bonaerenses, pp 863–866

Schalamuk I, Etcheverry R, Garrido L, Fernández R (1992) Geología y características tecnológicas de los depósitos de arcilla de los partidos de Azul y Lobería, provincia de Buenos Aires. In: Proc 4° Congreso Nacional y 1° Congreso Latinoamericano de Geología Económica, pp 477–488

Teruggi ME, Kilmurray JO (1975) Tandilia. Relatorio Geología de la Provincia de Buenos Aires. In: Proc 6° Congreso Geológico Argentino, pp 55–78

Teruggi M, Kilmurray J (1980) Sierras Septentrionales de la provincia de Buenos Aires. In: Turner JCM (ed) In: Proc 2° Simposio Geología Regional Argentina. Academia Nacional de Ciencias de Córdoba, Córdoba, Argentina II, pp 919–956

# Chapter 3
# Benito Juárez County

**Abstract** In the Benito Juárez County, four sectors will be considered: El Ferrugo and Constante 10-El Cañón Sector; Villa Cacique Sector; Sierra La Juanita Sector and Cuchilla de Las Aguilas-Sierra de La Tinta Sector. Crystalline basement rocks have been altered by weathering processes, resulting, from bottom upwards in: bedrock, saprock, saprolite and, occasionally, in two superimposed paleosols. Argillized basement rocks are covered by a highly resistant conglomerate of the Balcarce Formation. Weathering profiles are analized in detail. Mineralogical composition, by X-ray diffraction of the clays of El Ferrugo and Constante 10 is similar. Also, these deposits are similar to those of La Verónica and Santa María, described in Chap. 2. According to the technological characteristics of the clays of El Ferrugo and Constante 10-El Cañón Sector they are classified as "Fire clays". In the Villa Cacique Sector the Olavarría Formation, followed by the Loma Negra Formation and overlaid by the Cerro Negro and the Balcarce Formation, are described. The clays of the Cerro Negro Formation are composed of detrital illite and diagenetic clay minerals. Chemical and technological analyses attest to low values of PCE. The clays are classified as varied clays (wide-ranging). At the Sierra La Juanita Sector, the Villa Mónica Formation overlies unconformably the crystalline basement rocks and has been exploited for the ceramic industry. In the last years the Villa Mónica Formation has been redefined as carbonate, mixed, both with quartz megacrystals, and hetherolitic facies; their origin is explained and a paragenetic sequence is proposed. MISS are described in siliciclastic and mixed facies of the Villa Mónica Formation. SEM of the clay deposits and paleoenvironmental conditions of the Villa Mónica Formation are discussed. The Villa Mónica Formation age is considered to be Riphean, on the basis of the type of stromatolites. Technologically, clays from the Villa Mónica Formation are classified as plastic clays. In the Cuchilla de Las Aguilas and Sierra de La Tinta Sector the sedimentary sequence overlying the basement rocks is represented by the Sierras Bayas Group covered by the Las Aguilas Formation and the latter, in turn, by the Balcarce Formation. Alunite provided a Middle Permian age according to K–Ar dating (telogenetic stage). MISS are described in the Las Aguilas Formation. Plastic clays, with refractory and semiplastic varieties, are used in red ceramic and cement industry.

© Springer International Publishing Switzerland 2016  27
P.E. Zalba et al., *Gondwana Industrial Clays*,
Springer Earth System Sciences, DOI 10.1007/978-3-319-39457-2_3

**Keywords** Residual deposits · Saprock · Saprolite · Paleosols · Carbonate ·
Mixed · Hetherolitic facies · Stromatolites · Quartz megacrystals · MISS ·
Alunite · Permian telogenesis · Paragenetic sequence · Mineralogy · SEM ·
Technology

## 3.1   El Ferrugo and Constante 10—El Cañón Sector

### 3.1.1   Residual Deposits: Characteristics, Mineralogical, and Chemical Composition

To the SE of the sites described in Chap. 2, the El Ferrugo clay deposits, at Estancia
Tres Lomas (11 km east of national route N° 3), and the Constante 10-El Cañón
clay deposits (4.3 km NW of the López railway station), on the NE slopes of Cerro
La Tortuga (the López railway station is 42 km to the SE of Chillar village) form
part of a series of outcrops aligned in direction NW–SE (Fig. 3.1).

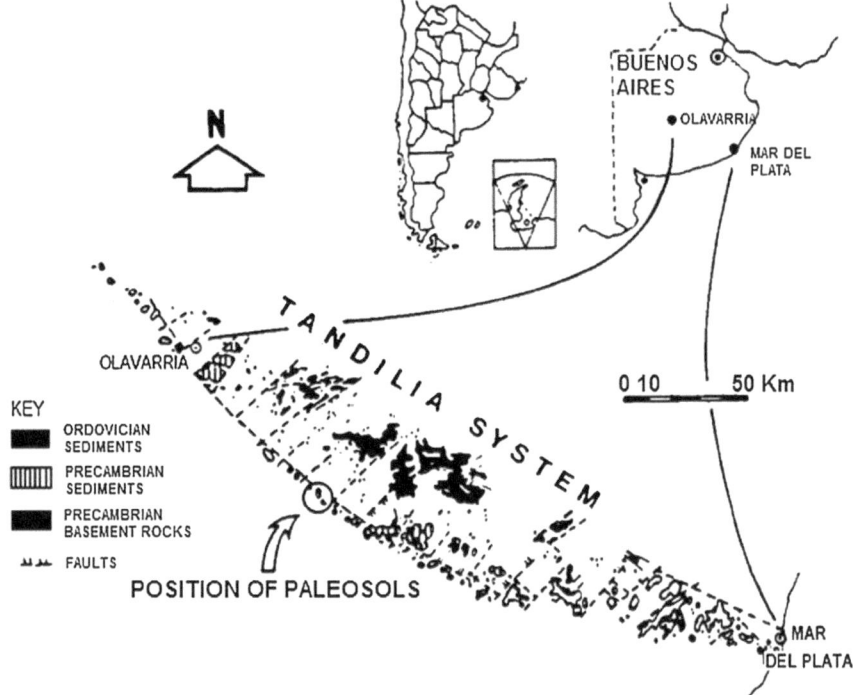

**Fig. 3.1**   Study area: outcrops of residual deposits. Taken from Iñiguez et al. (1990)

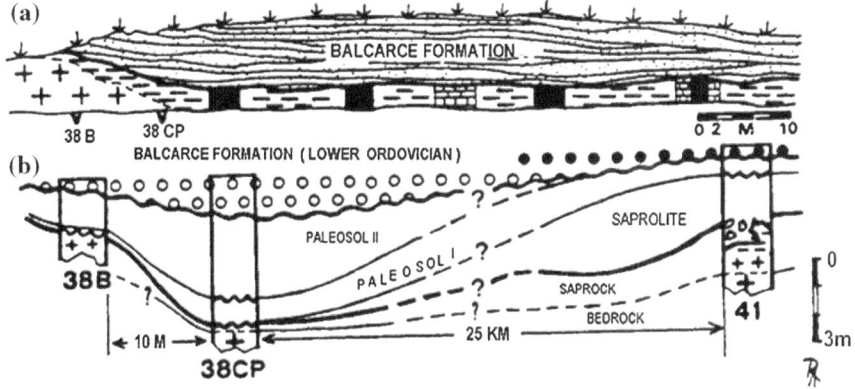

**Fig. 3.2** **a** Scheme of Constante 10 quarry showing the altered basement and quartzites of the Balcarce Formation. **b** Inferred reconstruction of the distribution of the bedrock, saprock, saprolite and paleosols, covered by Balcarce Formation. Taken from Iñíguez et al. (1990)

Crystalline basement rocks at Constante 10 outcrop have been altered by weathering processes, resulting from bottom upwards in: bedrock, saprock, saprolite, and two superimposed paleosols. Location of the study zone is shown in Fig. 3.2a. In Fig. 3.2b, the inferred reconstruction of the distribution of the different alteration zones of the basement and of the paleosols covered unconformably by the Balcarce Formation is offered. However, at El Cañón it is only possible to recognize bedrock and saprock zones, probably due to erosion processes related to a transgression occurred during the Early Ordovician, as indicated by Iñíguez et al. (1989).

In the El Ferrugo deposits, clays are essentially similar in thickness and mineralogical composition to those of the Constante 10-El Cañón and of La Verónica and Santa María outcrops, already described. In Constante 10 (exploited deposits), as the argillized basement rocks are covered by a thick (4 m), highly resistant conglomerate of the Balcarce Formation, it can be exploited in galleries, leaving pillars of material intact which support the roof of the mine, being one of the few sites of the Province of Buenos Aires exploited in this way.

Petrographic studies carried out in collaboration (Iñíguez et al. 1990) indicated that the bedrock of Constante 10 is fine grained, red and brown colored; composed of plagioclase (oligoclase: An 17–22 %), hornblende, monocrystalline quartz, micas, and scarce apatite. The biotites are deferrized and the resulting iron oxide released fills cracks and fissures. There is also scarce muscovite. Greenish-brown hornblende is partially altered and displays zircon inclusions. Irregular fissures cross the rock and are usually stuffed with kaolinite. Abundant smectite produced by the weathering of hornblende, as well as illite-smectite (I/S) and illite-vermiculite (I/V), constitute the fine fraction (Fig. 3.3a).

Transitionally (upwards), the saprock shows relics of the texture of the original rock, but most of the plagioclase has been transformed into kaolinite, while I/S still persists but smectite disappears. Muscovites are altered to I/S and quartz content

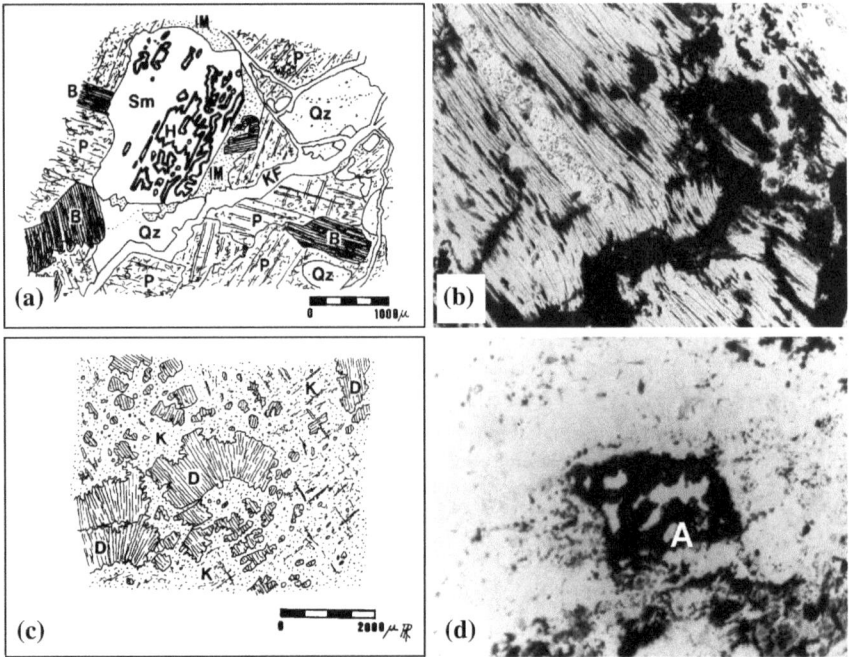

**Fig. 3.3** Petrographic studies of Constante 10 quarry. **a** Bedrock: abundant quartz (Qz), sericitized plagioclase (P), biotite (B), hornblende (H) altered to smectite (Sm). **b** Saprock: deferrized biotite (B) and Illitic material (IM) (160x magnification). **c** Paleosol II: vermiform crystal aggregates of dickite (D) among fine aggregates of kaolinite (K). **d** Paleosol I: crystals of diagenetic anatase, 100 microns in size, (40x magnification).Taken from Zalba and Andreis (2001)

decreases (Fig. 3.3b). The accessory minerals are zircon, titanite, and authigenic hematite and anatase. Abundant rosette aggregates of very fine-grained dickite have been characterized by infrared spectroscopy. Weathering of biotite led to the formation of vermiculite and of illite/vermiculite (I/V). Kaolinite associated with hematite and goethite are found filling cracks and fissures.

Upward, on the weathering profile a completely argillized zone is recognized: the saprolite. It is formed by abundant kaolinite-dickite and, to a lesser extent, by I/S. A few crystals of plagioclase, still distinguishable externally, have fully turned into kaolinite. Micas have also transformed completely into kaolinite. Quartz varies between 20 and 30 %, and decreases upwards.

In some areas up to two paleosol levels, product of the remobilization of underlying material, as well as sedimentary input, have been detected. They are composed of abundant dickite vermiform aggregates in a fine matrix of kaolinite (Fig. 3.3c) and lower proportion of I/S. Biotite is fully deferrized. Anatase (Fig. 3.3d) and titanite are common. Cracks and fissures are filled with kaolinite and iron hydroxides (goethite). The effect of elutriation (transport of solid materials from higher levels in the profile and its deposition along fissures and cracks) is responsible for the filling of fractures with kaolinite and iron oxides and hydroxides at all levels of the stratigraphic profile; excellent indication that it is a weathering profile.

According to petrographic studies carried out, the rocks of the crystalline basement of the El Ferrugo and Constant 10 are very similar to those of the La Verónica. The rocks are classified, in this case, as biotitic tonalities and as tonalites.

The El Ferrugo clays are whitish in color and, in Constante 10, whitish to greenish-reddish at the bottom due to the increase, in depth, of iron oxides (hematite). The rocks are massive, well compacted, with "eyes" of greenish material spread unevenly and variable amount of quartz. The mineralogical composition by X-ray diffraction of the clays of El Ferrugo and Constante 10 is similar, and can be consulted in Table 3.1. Kaolinite is the most abundant clay mineral, followed by I/S.

The chemical composition of major elements of the clays of El Ferrugo and Constante 10 is shown in Table 3.2. Variations in the percentages are attributed to the fact that samples from El Ferrugo correspond to saprock levels, while samples from Constante 10 belongs to saprolite levels. Indeed, kaolinite (with dickite) is more abundant in Constante 10 but illite-smectite and micas are scarce. As the clays have formed "in situ", and are rich in kaolinite, a more advanced degree of transformation of the basement rock feldspars in kaolinitic clay minerals has been

**Table 3.1** Mineralogical composition by X-ray diffraction of the clays of El Ferrugo and Constante 10

| Mineralogical composition | El Ferrugo | Constante 10 |
|---|---|---|
| Kaolinite (%) | 63 | 70 |
| Illite-smectite and/or muscovite (%) | 6 | 21 |
| Quartz (%) | 31 | 9 |

Taken from Garrido et al. (1984)

**Table 3.2** Chemical composition of the clays of El Ferrugo and Constante 10

| Sample Weigth % | El Ferrugo | Constante 10 |
|---|---|---|
| $SiO_2$ | 60.3 | 46.8 |
| $Al_2O_3$ | 25.5 | 36.0 |
| $Fe_2O_3$ | 1.4 | 0.7 |
| $TiO_2$ | 1.4 | 1.5 |
| CaO | 2.2 | 1.0 |
| MgO | 0.6 | 0.3 |
| MnO | – | – |
| $Na_2O$ | 0.6 | 0.2 |
| $K_2O$ | 0.4 | 1.5 |
| $H_2O^-$ | – | – |
| $H_2O^+$ | 9.0 | 11.8 |

Taken from Garrido et al. (1984)

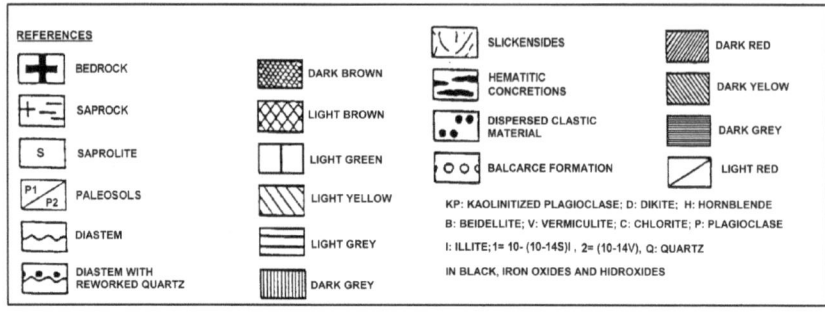

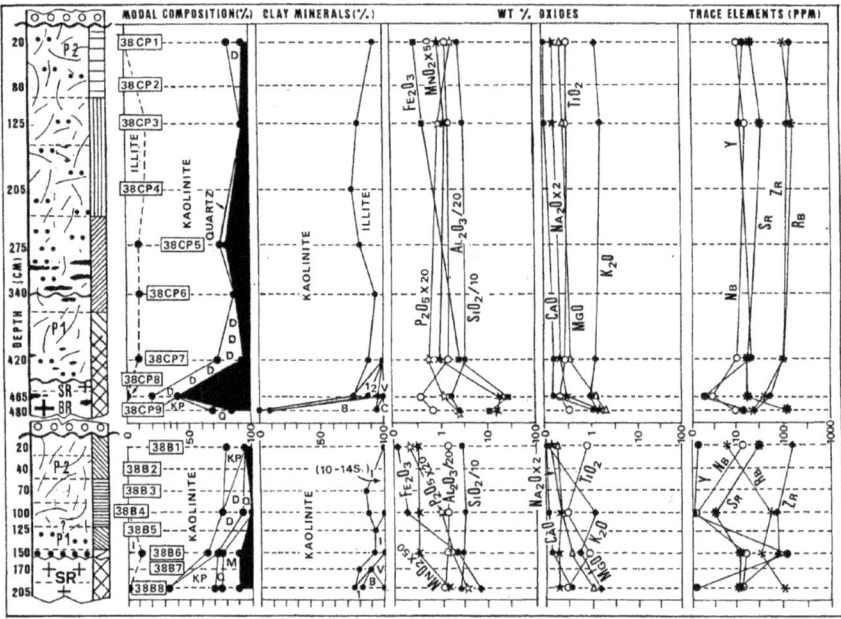

**Fig. 3.4** Profiles CP 38 and 38B. Constante 10. Weathering profiles, vertical mineral variations and chemical composition. Taken from Iñíguez et al. (1990)

reached and, at the same time, a major transformation of illite-smectite and micas, also in kaolinitic minerals, occurred towards the top of the deposit.

Detailed studies carried out in Constante 10 (Iñíguez et al. 1990) allowed the authors to corroborate the above affirmations. Figures 3.4 and 3.5 show three stratigraphic sections (P41, P38CP, and P38B), where it is possible to see modal analyses and vertical variation in the mineralogical and chemical composition and the separation, on the basis of diverse analyses, of different areas of weathering in the basement rock (bedrock, saprock, saprolite, and paleosols). Table 3.3 shows the variation in the percentages of clay minerals in the different areas of weathering.

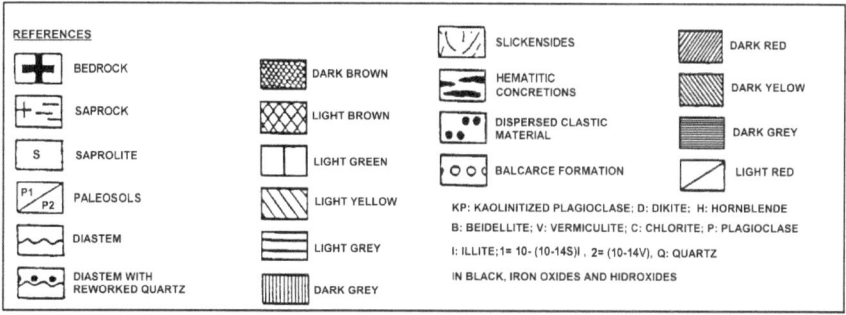

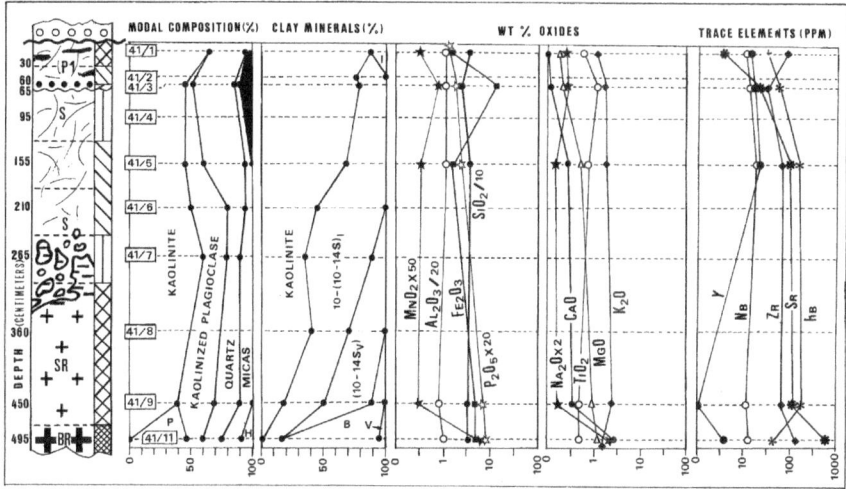

**Fig. 3.5** Profile P41. Constante 10. Weathering profile, vertical mineral variations and chemical composition. Taken from Iñíguez et al. (1990)

Table 3.4 shows the chemical analysis corresponding to these sections. P41, the most complete one, serves to summarize the variations found. Aluminum remains virtually immobile during weathering and was incorporated into the resulting clay minerals. Enrichment of aluminum in the paleosols, formed by weathering of the original rocks of the basement, is not pronounced as it would occur in other types of rocks that had a similar total composition. Its increase is interpreted as derivative of its remobilization from biotite, and subsequent incorporation to the clay minerals.

On the one hand, the silica is conservative, with some variations that attest to acidic weathering conditions. On the other hand, its increase in the paleosol 1 (P1) is consistent with the incorporation of detrital (sedimentary) components. The increase of titanium is interpreted as derived from the mobilization from biotite. Niobium and yttrium behave like titanium, and were presumably contained in titanium minerals and micas. Calcium and sodium, the most mobile elements, show losses of 99.76 and 89 % respectively when moving from the bedrock to the

**Table 3.3** Clay minerals variation (in percentage) of profiles P41, P38CP and P38B (Constante 10)

| Sampling depths (m) | Weathering zones | Sample | Kaolinite | 10-(10-14s) | Illite | (10-14v) | Montmorillonite | Vermiculite | Chlorite |
|---|---|---|---|---|---|---|---|---|---|
| 0.20 | P1 | 41/1 | 88 | – | 12 | – | – | – | – |
| 0.65 | S | 41/3 | 83 | – | – | – | – | – | – |
| 1.55 | | 41/5 | 67 | – | – | – | – | – | – |
| 3.60 | SR | 41/8 | 40 | 30 | – | 30 | – | – | – |
| 4.50 | | 41/9 | 18 | 32 | – | 41 | 10 | – | – |
| 4.95 | BR | 41/11 | – | 15 | – | – | 81 | 4 | – |
| 0.20 | P2 | 38CP1 | 95 | – | 5 | – | – | – | – |
| 1.25 | | 38CP3 | 83 | – | 17 | – | – | – | – |
| 4.20 | P1 | 38CP7 | 88* | – | 12 | – | – | – | – |
| 4.65 | SR | 38CP8 | 74* | 12 | – | 6 | – | 8 | – |
| 4.80 | BR | 38CP9 | – | 5 | – | – | 84 | 4 | 7 |
| 0.20 | P2 | 38B1 | 100 | – | – | – | – | – | – |
| 0.70 | | 38B3 | 88 | 12 | – | – | – | – | – |
| 1.00 | | 38B4 | 86* | 14 | – | – | – | – | – |
| 1.50 | P1 | 38B6 | 92 | – | 8 | – | – | – | – |
| 1.70 | SR | 38B7 | 79 | – | 9 | – | – | 12 | – |
| 2.00 | | 38B8 | 70 | 11 | 9 | – | 19 | – | – |

X-ray diffraction of oriented sample (fraction <2 microns). BD: bedrock, SR: saprock; S: saprolite; P1: paleosol 1; P2: paleosol 2. * Dickite also present. IR: expandability parameter. Taken from Iñiguez et al. (1990)

**Table 3.4** Chemical composition of profiles P41, P38CP and P38B (Constante 10) Zr, Nb, Y and Sr are expressed in ppm (parts per million)

| Sample | $SiO_2$ | $TiO_2$ | $Al_2O_3$ | $Fe_2O_3$ | MnO | MgO | CaO | $Na_2O$ | $K_2O$ | $P_2O_5$ | $H_2O$ | Total | Zr | Nb | Rb | Y | Sr |
|---|---|---|---|---|---|---|---|---|---|---|---|---|---|---|---|---|---|
| 41/1 | 60.96 | 0.80 | 24.25 | 1.48 | 0.01 | 0.28 | 0.01 | 0.20 | 1.30 | 0.07 | 10.28 | 99.64 | 95 | 15 | 45 | 15 | 5 |
| 41/3 | 46.38 | 1.03 | 23.15 | 17.40 | 0.02 | 0.33 | 0.02 | 0.18 | 1.66 | 0.16 | 9.53 | 99.86 | 40 | 17 | 72 | 20 | 36 |
| 41/5 | 57.04 | 0.88 | 22.85 | 1.38 | 0.01 | 0.74 | 0.41 | 0.10 | 2.55 | 0.19 | 12.9 | 99.05 | 92 | 29 | 135 | 36 | 104 |
| 41/9 | 54.17 | 0.63 | 18.60 | 6.36 | 0.01 | 0.98 | 0.56 | 0.10 | 3.72 | 0.44 | 13.98 | 99.55 | 75 | 10 | 218 | – | 158 |
| 41/11 | 52.75 | 0.65 | 19.85 | 6.26 | 0.15 | 1.82 | 4.18 | 1.80 | 1.84 | 0.47 | 9.8 | 99.58 | 178 | 13 | 64 | 6 | 850 |
| 38CP1 | 44.94 | 0.55 | 35.90 | 0.48 | 0.02 | 0.31 | 0.01 | 0.10 | 2.10 | 0.11 | 15.15 | 99.57 | 208 | 11 | 100 | 20 | 32 |
| 38CP3 | 45.40 | 0.55 | 36.50 | 0.63 | 0.03 | 0.41 | 0.01 | 0.10 | 3.36 | 0.06 | 12.58 | 99.63 | 130 | 25 | 168 | 12 | 54 |
| 38CP7 | 46.25 | 0.50 | 31.50 | 3.39 | 0.02 | 0.58 | 0.17 | 0.13 | 2.28 | 0.04 | 14.75 | 99.62 | 100 | 14 | 108 | 25 | 25 |
| 38CP8 | 22.85 | 0.29 | 12.05 | 46.50 | 0.65 | 1.20 | 0.11 | 0.21 | 1.12 | 0.07 | 14.95 | 100.0 | 62 | 5 | 56 | 2 | 24 |
| 38CP9 | 47.30 | 0.59 | 16.65 | 12.88 | 0.40 | 2.50 | 1.47 | 1.12 | 2.02 | 0.24 | 14.23 | 99.50 | 34 | 10 | 32 | 25 | 122 |
| 38B1 | 46.85 | 0.91 | 35.90 | 0.08 | 0.01 | 0.09 | 0.01 | 0.10 | 0.01 | 0.02 | 14.10 | 99.09 | 222 | 26 | 8 | 2 | 56 |
| 38B4 | 53.08 | 0.42 | 32.50 | 0.33 | 0.01 | 0.25 | 0.01 | 0.12 | 1.39 | 0.05 | 11.38 | 99.54 | 88 | – | 81 | – | 6 |
| 38B6 | 48.30 | 0.97 | 31.55 | 2.99 | 0.01 | 0.58 | 0.05 | 0.10 | 0.76 | 0.08 | 14.23 | 99.62 | 90 | 20 | 54 | 150 | 16 |
| 38B8 | 48.20 | 0.46 | 28.10 | 8.62 | 0.03 | 1.28 | 0.56 | 0.10 | 1.57 | 0.27 | 10.40 | 99.59 | 15 | 12 | 112 | 2 | 14 |

All other concentrations are in percent. Taken from Iñiguez et al. (1990)

paleosols. Large losses of these elements mark the base of the chemical weathering of the section, and the boundary between the bedrock and the saprock. Up in the profile, pronounced loss in sodium is the most consistent characteristic of paleosols of all ages. The loss of magnesium is consistent with the destruction of the micas and the formation of kaolinite-dickite. This suggests an intense process of decomposition that took place with enough water in the pores of the rock to allow the removal of this element, together with calcium and sodium, and also indicates a humid climate during weathering. Iron decreases from bedrock upward, but increases sharply at the base of the paleosol. During the formation of the saprolite, solutions must have been acidic and, when the paleosol was formed, there must have been a better oxygenation and the precipitation of iron took place. The behavior of iron is similar to manganese, except for the top of the profile (paleosol). There is a corresponding loss, although it is greater for manganese, as it should be expected if iron had been removed by weathering processes.

As potassium and rubidium are abundant in deep samples of the bedrock and saprock, and diminish in the paleosol and, taking into account that potassium feldspars have not been found, these elements must have been contained in the micas, and their decrease is consistent with their destruction towards the upper levels. Also, early diagenesis may be responsible for the increase in potassium in the saprock and in the saprolite given by the formation of I/S. Phosphorus loss is consistent with the absence of organisms able to retain it on the ground. The behavior of zirconium, which increases, generally upward is expected due to its high stability during weathering. The significant increase of phosphorus in the paleosol is a normal phenomenon. Finally the strontium is lost gradually upwards, consistent with the great mobility of this chemical element during weathering.

As for the Profile P38CP, greater enrichment of aluminum observed is related to a higher content in kaolinite-dickite in the paleosols of this section (P1 and P2). Titanium is preserved, except in the transition between the bedrock and the saprock. The loss of silicon between P1 and P2 coincides with a high concentration of iron because, as weathering processes reached this depth, indicated by the content of sodium, iron concentration must have been caused by remobilization and diagenetic precipitation. Erratic tendencies of some of the oxides in the Profile P38B are interpreted as caused by the existence of two successive paleosols, which under-went erosional processes and sedimentary input as well. These data represent a summary of the interpretations published in Iñíguez et al. (1990).

### 3.1.2   Scanning Electron Microscopy of Residual Clays of Constante 10

Scanning electron microscopy of clays from the saprock of Constante 10, Profile 41 (sample 41/8), shows three electron micrographs that allowed the authors to observe: an aggregate of smectite (probably with I/S), with "honeycomb" texture (Fig. 3.6a); pyramidal quartz with diagenetic dissolution associated to cementation

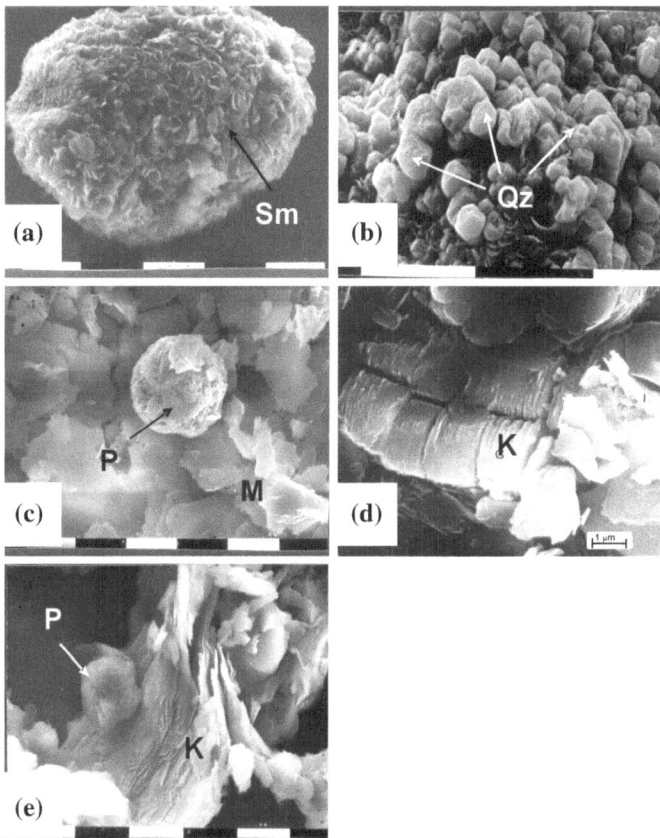

**Fig. 3.6** Scanning electron micrographs. Constante 10, Profile 41. Saprock (sample 41/8): **a** Aggregate of smectite (Sm), with "honeycomb" texture. **b** Pyramidal quartz (Qz) with diagenetic dissolution. **c** Pellet (P) of diagenetic $Fe_2O_3$ on mica (M) crystals. Saprolite (sample 41/8, paleosol 1): **d** Clay aggregates, predominantly kaolinite (K) and $Fe_2O_3$ pellet (P). Scale bar: **a** 10 μm, **b** 5 μm, **c** and **e** 1 μm. Taken from Iñiguez et al. (1990)

processes (Fig. 3.6b), and a pellet of diagenetic $Fe_2O_3$ on mica crystals (Fig. 3.6c). Already in the saprolite, kaolinite "accordions", formed from mica, are clearly visible (Fig. 3.6c, d). Also in Profile 41 (sample 41/1; paleosol 1), clay aggregates of diagenetic kaolinite, and a pellet of $Fe_2O_3$, are observed (Fig. 3.6e).

### 3.1.3   Technological Properties of Residual Clays of the El Ferrugo and Constante 10

It is noteworthy that in the whitish clays of Constante 10, dickite (mineral of the Kaolinite Group, of higher temperature than kaolinite) reaches up to 10 %, and probably is distributed in the above-mentioned "eyes". This mineral increases the refractory properties of the clays. The standard techniques used for measuring the technological properties are: Pyrometric Cone Equivalent (PCE), permanent linear change by heating, specific apparent weight, apparent porosity, water absorption, room temperature flexural strength in 3-point bending (MOR), and plasticity. From analysis of Table 3.5—which relates the number of milliequivalents exchanged per 100 g of clay (T value) with the plasticity index and the particle size distribution of the material—a correlation among the properties above mentioned is observed. In general, the T value obtained corresponds to the one expected for kaolinitic materials. In the El Ferrugo and Constante 10 clays (LGX 1032), with high percentage of particles larger than 74 microns, the exchange capacity is low showing a low plasticity index.

The results obtained from technological tests for different temperatures of burning are detailed in Tables 3.6 and 3.7 for each site and were carried out on specimens in semi-dry and plastic mud state. Heat treated (1200–1300 °C) clays of the El Ferrugo specimens (PCE 29) have very good appearance, without cracks or fissures, few signs of vitrification, and porosity around 20 %. When the sample is heated at temperatures above 1500 °C, the obtained specimens are very compact (1.6 % porosity), hard (high modulus of rupture), and with little shrinkage due to the high content of quartz in this material.

**Table 3.5** Relationship between the cation exchange capacity (T value), the plasticity index and the particle size distribution

| Sample | T value | Plasticity index | Fraction >74 $\mu$ (%) | Fraction <2 $\mu$ (%) |
|---|---|---|---|---|
| El Ferrugo | 3.8 | 3.9 | 13.0 | 29.6 |
| LGX 1032 | 5.8 | 6.6 | 9.6 | 35.0 |

Taken from Garrido et al. (1984)

**Table 3.6** Technological tests: El Ferrugo clay

| | Temperature (°C) | Linear shrinkage (%) | Apparent porosity (%) | Water absortion (%) | Apparent density ($10^{-3}$ kg./m$^3$) | MOR (Kpa) |
|---|---|---|---|---|---|---|
| Semi-dry state | 105–110 | – | – | – | – | 67.6 |
| | 1230 | 1.1 | 26.6 | 13.3 | 2.0 | 252.7 |
| | 1300 | 1.1 | 27.3 | 13.3 | 2.0 | – |
| | 1400 | 5.4 | 27.6 | 14.0 | 1.9 | 342.1 |
| | 1500 | 5.8 | 1.6 | 0.7 | 2.4 | 1394.4 |
| | 1600 | 2.6 | 1.0 | 0.4 | 2.3 | 961.5 |
| Plastic mud state | 105–110 | 3.5 | – | – | – | – |
| | 1230 | 5.7 | 37.7 | 22.6 | 1.7 | – |
| | 1300 | 6.4 | 36.2 | 21.3 | 1.7 | 63.7 |

Color after calcination at 1230 °C: white (2.5 Y 8/2 Munsell soil color chart). Refractoriness: PCE 29 (1650 °C). Taken from Garrido et al. (1984)

**Table 3.7** Technological tests: LG XT1032 clay (Constante 10)

|  | Temperature (°C) | Linear shrinkage (%) | Apparent porosity (%) | Water absorption (%) | Apparent density $(10^{-3\ kg}./m^3)$ | MOR (Kpa) |
|---|---|---|---|---|---|---|
| Semi-dry state | 105-110 | – | – | – | – | 41.22 |
|  | 1230 | 3.2 | 19.5 | 9.1 | 2.1 | 723.3 |
|  | 1300 | 5.8 | 12.9 | 4.6 | 2.8 | 749.9 |
|  | 1400 | 6.4 | 5.5 | 2.3 | 2.4 | 1665.5 |
|  | 1500 | 6.4 | 0.3 | 0.6 | 2.4 | 1436.7 |
|  | 1600 | 4.7 | 1.0 | 0.4 | 2.3 | 1070.7 |
| Plastic mud state | 105-110 | 4.3 | – | – | – | – |
|  | 1230 | 6.4 | 32.2 | 19.1 | 1.7 | – |
|  | 1300 | 8.3 | 24.2 | 13.3 | 1.8 | 182.6 |

Color after calcination at 1230 °C: white (10 YR 8/2 Munsell soil color chart). Refractoriness: PCE 31 (1683 °C). Taken from Garrido et al. (1984)

Constante 10 (LGX 1032) clay is refractory, which corresponds to a PCE 31 and is suitable for the manufacture of highly refractory parts. Semi-dry, pressed specimens present good "green resistance". They are calcined up to 1600 °C. At 1230 ° C it shows little shrinkage and porosity of 19 %, while at temperatures above 1400 °C, pieces sintered with porosity between 0.3 and 1 %, although without deformations or cracks and high mechanical resistance. The analysis of the mineralogical, chemical, physical, and ceramic characteristics sets that the El Ferrugo clays, for its refractoriness, linked with the high percentage of kaolinite and convenient clear color are suitable for the manufacture of refractories of medium quality. According to the characteristics of the clays of El Ferrugo and Constante 10 they are classified as "Fire clays".

### 3.1.4   The Villa Cacique, La Juanita, Cuchilla de Las Aguilas, and Sierra de La Tinta Sectors

Close to the town of Barker, about 60 km to the SW of the city of Tandil, in the Benito Juárez County, clay sedimentary deposits of industrial importance, corresponding to different geological formations of different ages, are situated. These formations extend from NW to SE along Villa Cacique, La Juanita, Cuchilla de Las Aguilas, and Sierra de La Tinta localities having been detected by drilling and by the opening of numerous quarries, some of them currently in exploitation. From base to roof (see Table 1.1) the following formations have been identified in the above-mentioned areas: Sierras Bayas Group (comprising: the La Juanita Formation, the Cerro Largo Formation, the Olavarría Formation, and the Loma Negra Formation); the Cerro Negro Formation and the Las Aguilas Formation (all Neoproterozoic), and, finally, the Balcarce Formation (Ordovician).

Next, each of these sectors will be addressed separately.

## 3.2   Villa Cacique Sector

### 3.2.1   *Sedimentary Deposits; Characteristics, Mineralogical, and Chemical Composition*

At Villa Cacique, 5 km to the SW of the town of Barker, the quarries open for limestone exploitation by the Loma Negra CIASA (Barker) cement factory have left part of the sedimentary sequence uncovered: limestones and claystones (Fig. 3.7). In the middle of the twentieth century, the Loma Negra cement factory was located in Barker, working at first with a single furnace (1955/56). By 1961, a second furnace was incorporated. The Barker factory had its boom in the decades of the 1960s and 1970s and closed its doors in 2001. Currently the Loma Negra limestone deposits were acquired by the Camargo Correa Cementos SA, a Brazilian capital group (2005).

The limestone, 25 m thick, which varies from black to brownish in color, bears interbedded illitic clays and has been used in the Province of Buenos Aires cement industry in recent years. This limestone corresponds to the Loma Negra Formation (see Table 1.1). The base of the quarries, product of the Loma Negra Formation limestone exploitation, corresponds to the top of the Olavarría Formation (see Table 1.1), consisting of turquoise to light greenish, laminated, illitic clays, with

**Fig. 3.7**   Loma Negra limestone quarry, Villa Cacique, showing the following Neoproterozoic units: Loma Negra Formation (*1*); Cerro Negro Formation (*2A*-quartzite channeled and *2B*-clay sediments); Balcarce Formation (*3*) and Quaternary deposits (*4*). Taken from Zalba and Andreis (2001)

very well developed, fresh, and visible pyrite, and impurities of calcite. These clays are not exploited in the area. From the structural point of view, these clays are important because they are found as a continuous, horizontal mud bed, and several connected pipes of different size and width which penetrate the overlying Loma Negra Formation in different areas of the Sierras Bayas locality. The Olavarría Formation clays will be described in detail in the Sierras Bayas Sector (Chap. 5) where the type locality is situated.

The Cerro Negro Formation clays (see Table 1.1) overlie, in erosive unconformity, the limestone of the Loma Negra Formation. With an exposed thickness of 15 m, the clays are olive gray, pink, ocher, and black, the latter found at the base and only detected by perforations in this sector. The Cerro Negro Formation is found either as collapsed deposits, produced by karstic dissolution of the underlying limestone roof or the filling of old existing caves whose roof sagged, or similarly by dissolution of the limestone occurred previously to the deposition of the Cerro Negro Formation, known in the geological literature as sinkholes (Zalba and Andreis 2001). These variations exist both in the Loma Negra quarry, Villa Cacique (Figs. 3.8 and 3.9) and in San José del Cármen quarry, located 12 km SE of Barker, 10 km SSE of Villa Cacique, and 10 km NE from the La Negra railway station. At this point, the political divisions of Juárez, Tandil, and Necochea counties converge. In the last county: Necochea, the San José del Cármen quarry is found. Deposits of up to 5 m in thickness of the Cerro Negro Formation have also been detected in a section of the road that leads from Barker to San José del Cármen, locally known as Lomada Blanca, 3 km NW from the latter.

**Fig. 3.8** Sinkholes (D) in *black* limestones of the Loma Negra Formation, filled with *red* clay deposits of the Cerro Negro Formation. Paleosurface (PS): in *white*. Villa Cacique. Taken from Zalba and Andreis (2001)

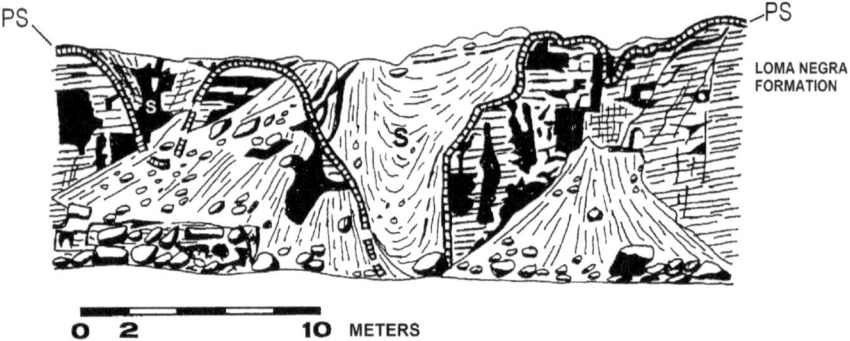

**O  2            1O** METERS

**Fig. 3.9** Scheme of sinkholes (D) in *black* limestones of the Loma Negra Formation, filled with *red* clay deposits of the Cerro Negro Formation. Villa Cacique. Paleosurface (PS). Taken from Zalba and Andreis (2001)

The Cerro Negro Formation defined for the first time for the area of Cerro Negro, Sierras Bayas (Iñíguez and Zalba 1974), considered as the stratotype area. It consists of more than 150 m, measured by drilling of red and gray shales overlying the limestone of the Loma Negra Formation. Seven years after that important discovery, Zalba (1981) located the same unit in the area of Villa Cacique (Barker) because of the mining work at the Loma Negra limestone quarry. The author correlated the new discovered deposits with those described first at the Cerro Negro locality based on geological and compositional criteria. Some years ago, this unit was also found at the subsoil of the basin—due to private companies drilling labor—in the same stratigraphical position as the one described by Zalba (1981) confirming this author's data.

The Cerro Negro Formation, at Villa Cacique, begins with a basal limestone breccia in a quarztitic matrix, with abundant fresh pyrite (Fig. 3.10). The sequence continues with channelized, lenticular quartzitic deposits, 2–5 m thick and, upward, with gray-olive lenticular claystones laminated deposits with intercalations of thin siltstone levels. The visible thickness of this sequence is up to 15 m (Fig. 3.11). It is covered, in erosive unconformity, by quartzite deposits bearing trace fossils which correspond to the Balcarce Formation (see Table 1.1). Recent deposits (loess and silts) of variable thickness cover the described sequence (upper deposits shown in Fig. 3.11).

The mineralogical composition of the clays of the Cerro Negro Formation in Villa Cacique, on a road cut between Barker and San José del Cármen quarry, and in San José del Cármen quarry, can be seen in Table 3.8. According to X-ray diffraction analyses, the clays are represented by illite-smectite (I/S), illite and chlorite-smectite (C/Sm), with abundant quartz and scarce feldspars as impurities. Also scarce smectite and alunite, on some levels, have been detected in Villa Cacique, while traces of halloysite and kaolinite have been identified in the upper part of the sequence. All these minerals, except illite, are attributed to diagenetic processes. Chemical analyses on samples of the Cerro Negro Formation at Villa Cacique can be seen in Table 3.9.

**Fig. 3.10** Basal limestone breccia in a quarztitic matrix, Cerro Negro Formation, Villa Cacique

### 3.2.2   Scanning Electron Microscopy of Clays of the Cerro Negro Formation at Villa Cacique, San José Del Cármen and Lomada Blanca

SEM shows the texture of the Cerro Negro Formation clays at Villa Cacique; Lomada Blanca and San José del Cármen localities. Texture shows face-to-face disposition of the clay minerals, typical of marine sedimentary deposition, and composed of chlorite-smectite. On these clays, pseudocubic crystals of diagenetic alunite are observed (Fig. 3.12a). Figure 3.12b shows a "honeycomb" texture of clays from Lomada Blanca consisting of diagenetic illite-smectite. Flakes are perpendicular to the surface of deposition. In Fig. 3.12c smectite fills cracks, formed by diagenetic processes much younger than most of the other diagenetic minerals.

Figure 3.12d represents the texture of essentially illitic clays from the same formation, at Calera San José del Cármen. Face-to-face disposition of laminae is an indication of deposition in a marine environment (detrital). The Cerro Negro

**Fig. 3.11** Gray-olive lenticular claystones, interbedded with siltstones of the Cerro Negro Formation (up to 15 m of visible thickness) lying unconformably on the Loma Negra Formation, Villa Cacique. Above, recent deposits

**Table 3.8** Different mineral association of clays of the Cerro Negro Formation in several deposits in the Villa Cacique sector

| | |
|---|---|
| Pinkish clays. Villa Cacique | Illite (ab), Montmorillonite (sc), Chlorite (tr) |
| Gray-olive clays. Villa Cacique | Illite, Chlorite (tr), Kaolinite (tr) |
| Clays filling limestones "San José del Cármen" | Illite (ab), Montmorillonite (sc) |
| Cut way to "San José del Cármen" | Illite (ab), Chlorite (ab), Interestratified Chlorite-Montmorillonite (sc) Illite, Kaolinite, Montmorillonite (sc) Illite (ab), Montmorillonite (sc), Kaolinite (sc) |
| Cerro Negro Formation | Chlorite (ab), Illite (ab), Interestratified Chlorite-Montmorillonite (sc) |

References: ab: abundant; sc: scarce; tr: trace. Taken from Zalba (1981)

Formation clays are exploited neither in Villa Cacique, nor in Calera San José del Cármen because of their little thickness.

### 3.2.3   Technological Properties of Clays of the Cerro Negro Formation at Villa Cacique

Technological analyses performed on the Cerro Negro Formation clays in Villa Cacique indicate that they possess Pyrometric Cone Equivalent (PCE) ranging between 13 and 14. These low values are due to the high content of alkalis and, in

**Table 3.9** Chemical composition of clays of the Cerro Negro Formation, Villa Cacique area

| Sample weigth %        | Olive-gray clays | Pinkish clays |
|------------------------|------------------|---------------|
| $SiO_2$                | 60.50            | 60.70         |
| $Al_2O_3$              | 21.80            | 22.20         |
| $Fe_2O_3$              | 4.50             | 4.70          |
| $TiO_2$                | 0.75             | 0.70          |
| CaO                    | 0.38             | 0.10          |
| MgO                    | 1.35             | 1.45          |
| $Na_2O$                | 0.65             | 0.50          |
| $K_2O$                 | 4.80             | 4.30          |
| $CO_2$ in $(CO_3Ca)$   | 0.50             | 0.00          |
| LOI                    | 5.23             | 5.35          |

Taken from Zalba (1981)

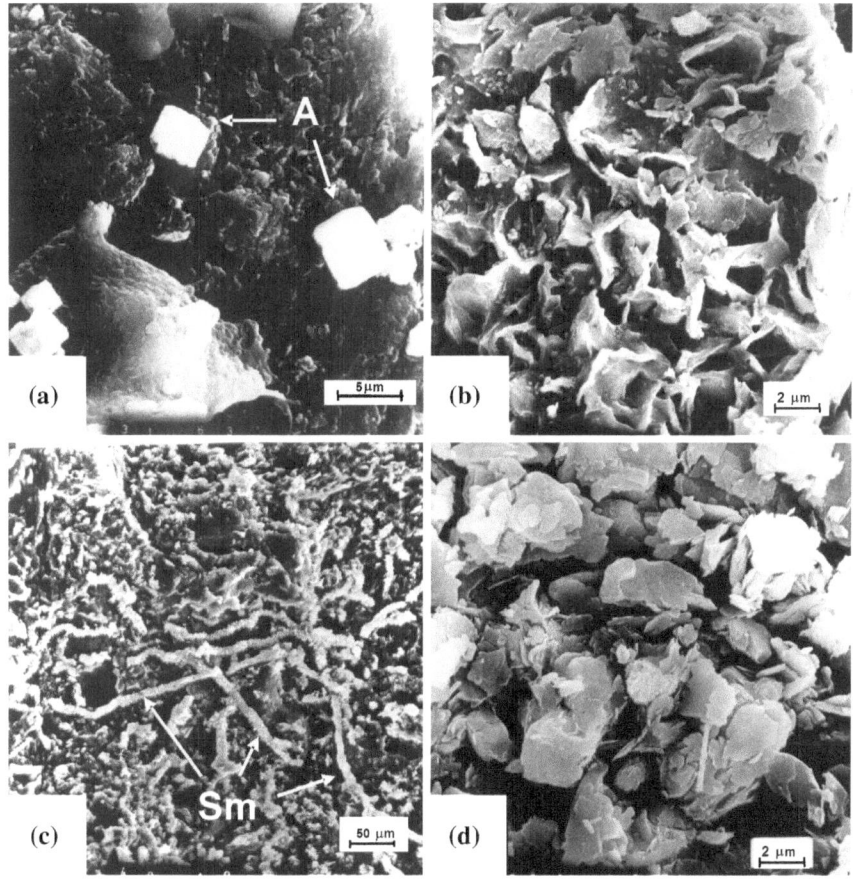

**Fig. 3.12** Scanning electron micrographs of the Cerro Negro Formation. Villa Cacique: **a** Chlorite-smectite clays with pseudo cubic crystals of diagenetic alunite (A). Lomada Blanca: **b** "Honeycomb" texture of illite-smectite clays. **c** Chlorite-smectite clays with diagenetic smectite (Sm) filling cracks. Calera San José del Cármen: **d** Illitic clays with face-to-face disposition of laminae (marine environment). Taken from Zalba (1988)

the absence of high-alumina minerals, the material has no refractory characteristics and its melting temperatures are low. In addition to this, the percentage of iron oxide also causes the decrease of the temperature of fusion (Zalba 1981).

The Cerro Negro Formation clays in Villa Cacique are classified as varied clays (wide-ranging) and its technological features will be given in detail when the deposits of the Cerro Negro locality type area (Sierras Bayas) are treated.

## 3.3  La Juanita Sector

### 3.3.1  Sedimentary Deposits; Included "in situ" Weathered Dolostones ("Ferruginous Clays")

At the Sierra La Juanita, the Villa Monica Formation deposits, the oldest sequence of the Sierras Bayas Group, (see Table 1.1) have been exploited for the ceramic industry, until few years ago, in fields of the Estancia La Siempre Verde, at La Placeres and Don Camilo quarries, located 4.5 km East; 14 km East and 20 km Southeast of the town of Barker, respectively. The Villa Monica Formation (locally also known as La Juanita Formation) overlies in erosive unconformity, grayish and weathered granitic crystalline basement rocks. It has been defined by Poiré (1987) at Sierras Bayas, where it was also named as Tofoletti sequence (Poiré and Spalletti 2005), with 52–70 m thick and bounded by unconformities at the base and on the roof of the unit as well.

In the same locality, Schauer and Venier (1967) were the first authors to recognize 10 m of dolostones at the Cerro de La Cruz, 2 km east from the Estancia La Siempre Verde. Since that date (over 40 years ago) these rocks were never found again until Zalba et al. (2007b) described them "in situ" in a very close area. About 20 years later of the Schauer and Venier (1967) discovery, Manassero (1986) described ferruginous clay associations between the Cuarcitas Inferiores and the Cuarcitas Superiores in a very close area to that described by Schauer and Venier (1967). In addition, Manassero (1986) found idiomorphic, pyramidal quartz crystals, of great development, although recognizing that they were never found "in situ".

Almost at the same time, Alló et al. (1986) also described presumably diagenetic quartz megacrystals which, according to the authors, were contained in "ferruginous clays". Alló (Unpublished thesis 2001), López et al. (2002) and López (Unpublished thesis 2006) also described in the Sierra La Juanita "ferruginous clays" yellowish-brown to yellowish colored, with reddish and green interbedded clays over the Cuarcitas Inferiores, as the unique lithology present in the Villa Mónica Formation. It should be noted that Alló (2001, unpublished thesis) described thin sections of dolostones, but she explained that she never found them "in situ" and, therefore, the dolostones could not be located in the stratigraphic sequence.

New stratigraphic sections in the Villa Mónica Formation were carried out in La Siempre Verde, La Placeres, and Don Camilo quarries (Zalba et al. 2010a) and integrated in a unique section, shown in Fig. 3.13, which allowed Zalba et al. (2010a) to redefine the existing lithologies overlying the Cuarcitas Inferiores, which now include: carbonate (C); mixed (carbonate/siliciclastic: C/S), and heterolithic (H) facies. C facies are represented by head, brownish, well-preserved columnar stromatolite dolostones. They are well exposed at the Estancia La Siempre Verde, whereas only the top of these deposits outcrops at La Placeres quarry and, at Don Camilo quarry they are not exposed. These rocks form bioherms (mounds) up to 4 m high (Fig. 3.14a), containing scarce sandy siliceous trapped sediments and forming cavernous structures. The mounds show distinctive columnar stromatolite structures (Fig. 3.14b) and are crossed by fractures, filled with silica or with calcite cement. Random aggregates of pyramidal quartz megacrystals, up to 20 cm long and 10 cm wide, are found growing in fenestral cavities in the dolostones (Fig. 3.14c, d).

The C/S facies show up to 6 m thick banded, brownish to yellowish weathered dolostones (Fig. 3.15a), with intercalated siliciclastics, the latter represented by greenish, laminated clay, silt, or sand beds. The weathered dolostones also bear loose, individual, or random aggregates of pyramidal quartz megacrystals ranging from approximately 1–10 cm long and 5 cm to few mm wide (Fig. 3.15b). Complex fracture-network systems, filled with reddish clays showing slickensides, cut the siliciclastic deposits and the weathered dolostones in several directions at different levels (Fig. 3.15c), and also have spread along sedimentary discontinuities (e.g., between the C/S and the upper H facies). Quartz megacrystals, up to 20 cm long were found "in situ" for the first time (Zalba et al. 2010a) in fresh dolostones as aggregates as well as individual pyramidal crystals (Fig. 3.15d).

According to Zalba et al. (2010a), the H facies, with a maximum thickness of 2 m, consist of a thickening and coarsening upwards, cross-laminated and rippled lenticular quartzites with intercalated greenish clay deposits (Fig. 3.16a, b)

### 3.3.2 Characteristics, Mineralogical, and Chemical Composition

Mineralogical X-ray diffraction data on samples of the Estancia La Siempre Verde, La Placeres, and Milli quarries can be found in Table 3.10 and have been taken from Alló (2001). The mineralogy of samples from Table 3.10 corresponds to: (1) detrital illitic clays (yellow and green colored) intercalated in the weathered dolostones, and (2) clay introduced in fractures, cracks, and available pore space (white, red colored). Chemical analysis of samples in which X-ray diffraction was carried out are presented in Table 3.11, according to data from Alló (2001).

As stated by Zalba et al. (2010a) optical microscopy shows textures of the carbonate (C); mixed: carbonate/siliciclastic (C/S), and Heterolithic (H) facies. Carbonate facies (C): it consists of fresh dolostones with up to 79 % dolomite, less

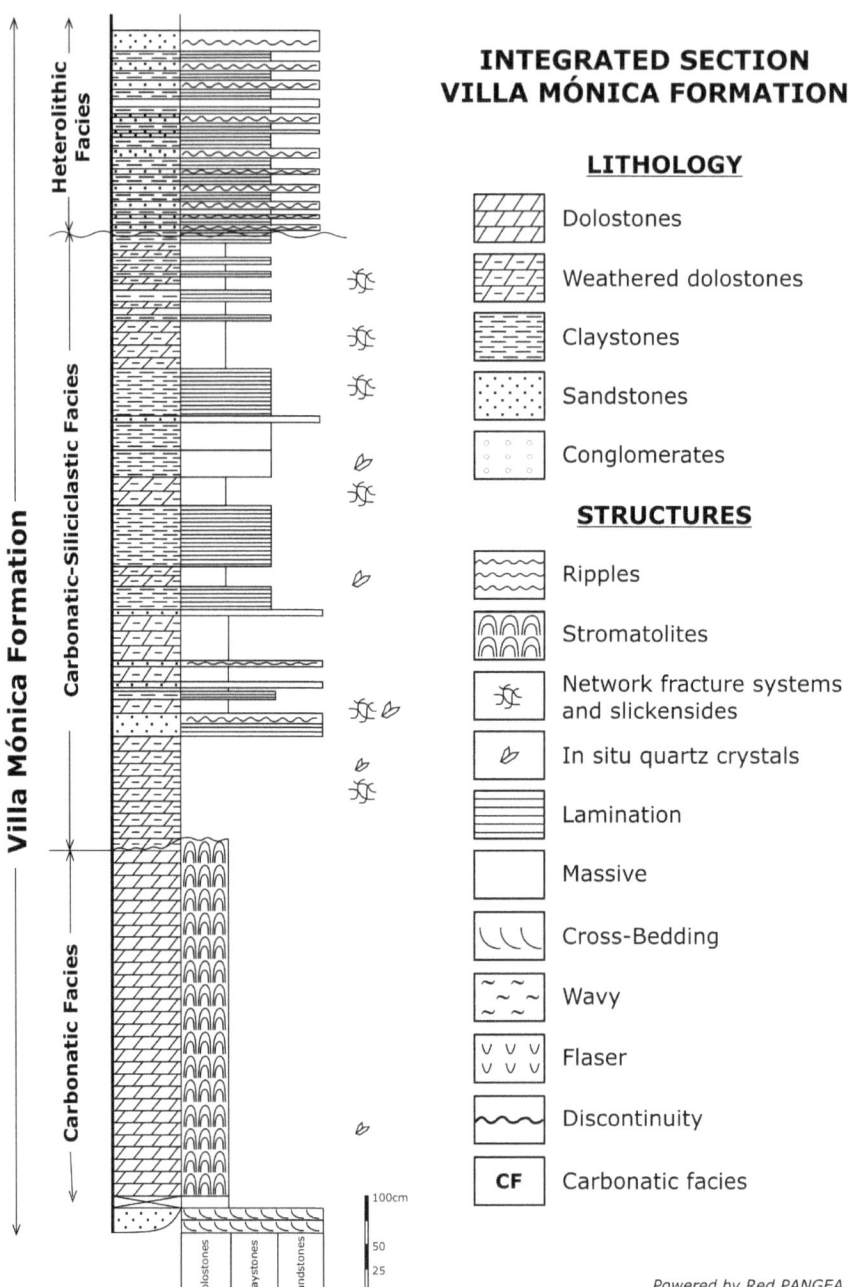

**Fig. 3.13** Integrated stratigraphic section of the Villa Mónica Formation, Sierra La Juanita. Taken from Zalba et al. (2010a)

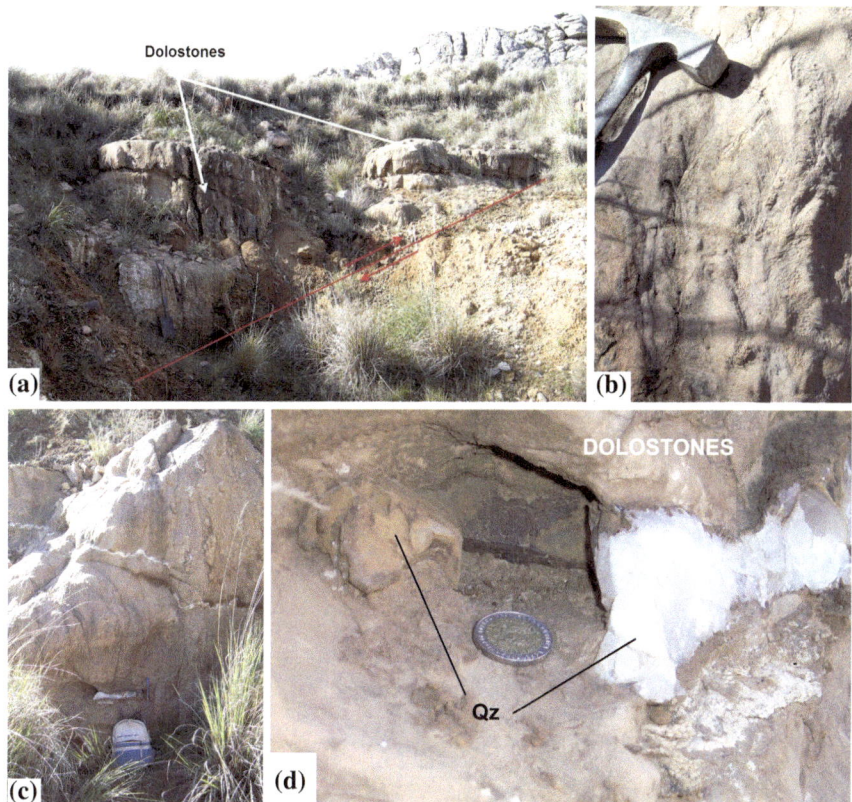

**Fig. 3.14** **a** Carbonate facies (fresh dolostones): General view of stromatolite bodies (mounds) of the Villa Mónica Formation, Sierra La Juanita. Taken from Zalba et al. (2010a) **b** Carbonate facies: detail of columnar stromatolite structures. Taken from Zalba et al. (2010a) **c** Quartz megacrystals growing in fenestral cavities in the dolostones, Villa Monica Formation, Estancia La Siempre Verde, Sierra La Juanita. Taken from Zalba et al. (2010a) **d** Pyramidal quartz megacrystals, found in weathered dolostones (carbonate-siliciclastic facies) Villa Monica Formation, Estancia La Siempre Verde, Sierra La Juanita. Taken from Zalba et al. (2010a)

than 10 % calcite and impurities of illite, goethite, and quartz. The fabric is represented by bimodal rhombohedric carbonate crystals of <340 and <250 μm in size, respectively (Fig. 3.17a, b). Contacts between grains are planar. Staining with Alizarin red shows that dolomitization was complete and that sparry calcite is present in fractures and in very small cavities (Fig. 3.17c). The rocks show typical stromatolite lamination (Fig. 3.17d) which attests to the precipitation of calcite from algae activity. On the basis of Dunham (1962), classification these rocks are columnar boundstones.

Epitaxial calcite rims on dolomite crystals are observed, being even and uniform in thickness (45 μm) and with lengths of 300 μm. Stained calcite rims (Fig. 3.17e, red in the photograph) are easily distinguishable from dolomite crystals. Ferric oxides and hydroxides are represented by hematite and abundant goethite, with

**Fig. 3.15 a** Carbonate-siliciclastic facies (C/S): thick banded weathered dolostones with siliciclastic intercalations, Villa Monica Formation, Estancia La Siempre Verde, Sierra La Juanita. Taken from Zalba et al. (2010a) **b** Individual or random aggregates of pyramidal quartz megacrystals dispersed in the weathered dolostones, Taken from Zalba et al. (2010a) **c** Siliciclastic facies, Villa Monica Formation. A complex fracture-network system filled with reddish clays, with slickensides cutting the weathered dolostones in several directions at different levels. Taken from Zalba et al. (2010a)

**Fig. 3.16 a** General view of the heterolithic facies of the Villa Monica Formation, Sierra La Juanita area. Taken from Zalba et al. (2010a) **b** Detail view of the heterolithic facies outcrop. Taken from Zalba et al. (2010a)

**Table 3.10** Mineralogical composition by X-ray diffraction of yellow, green, red and white samples of La Siempre Verde, (LSV), La Placeres (LP) and Milli quarries

| Quarry | Sample | Smectite | Illite | Kaolinite | Quartz | Hematite |
|--------|--------|----------|--------|-----------|--------|----------|
| LSV | 1 yellow | | 51 | 11 | 18 | 20 |
| LSV | 2 yellow | | 75 | 4 | 13 | 8 |
| LSV | 3 yellow | | 66 | 6 | 13 | 15 |
| LSV | 4 yellow | | 74 | 4 | 9 | 13 |
| LSV | 5 yellow | | 67 | 3 | 12 | 18 |
| LSV | 8 green | | 90 | | 10 | |
| LSV | 9 red | | 24 | 40 | 3 | 33 |
| LSV | 7 red | 15 | 20 | 33 | 5 | 27 |
| LP | 12 yellow | | 69 | | 22 | 9 |
| LP | 14 yellow | | 79 | | 14 | 7 |
| LP | 15 green | | 83 | | 17 | |
| LP | 20 red | | 14 | 74 | 3 | 8 |
| LP | 11 white | 23 | | 73 | 4 | |
| LP | 21 white | | 41 | 57 | 2 | |
| LP | 13 yellow | | 76 | | 24 | |
| LP | 22 white | | 28 | | 72 | |
| Milli | 27 yellow | | 46 | | 17 | 37 |
| Milli | 31 yellow | | 76 | | 18 | 6 |
| Milli | 34 yellow | | 78 | | 14 | 8 |
| Milli | 28 green | | 76 | | 24 | |
| Milli | 32 green | | 65 | | 35 | |
| Milli | 26 red | 18 | 19 | 43 | 3 | 17 |
| Milli | 33 red | | | 85 | 3 | 12 |

Taken from Alló (2001)

minor ferric oxides also present in fractures and surrounding dolomite crystals having precipitated prior to the epitaxial calcite rims.

The dolostones show silicification processes (Fig. 3.17f). Quartz is present as individual crystals or as random aggregates of megacrystals, which have grown in major cavities, in fractures or as pseudomorphs after rhombohedric dolomite crystals, also surrounded by hematite and/or goethite. Small cavities and fractures are filled with calcite or with quartz. X-ray diffraction analyses show that they consist of 79 % replacement dolomite, quartz, and illitic material (ISII) with <15 % of expansive layers (Table 3.12, sample 00).

Carbonate/siliciclastic facies (C/S): The carbonates of this facies are represented by relic domal or laminar stromatolite dolostones, where rhombohedric dolomite crystals have lost their continuity and have been completely replaced by illitic material (ISII), with less than 15 % of expansive layers (Table 3.12). The illitic clay includes rutile needles, which are 20–30 μm long, intersecting to form asterisk-shaped units showing a sagenite-like texture (Fig. 3.18a). Some of the relic

**Table 3.11** Chemical composition of yellow, green, red and white samples of La Siempre Verde, (LSV), La Placeres (LP) and Milli quarries

| Quarry | Sample | SiO₂ | Al₂O₃ | Fe₂O₃ | MnO₂ | MgO | CaO | Na₂O | K₂O | TiO₂ | P₂O₅ | LOI |
|---|---|---|---|---|---|---|---|---|---|---|---|---|
| LSV | 1 yellow | 44.86 | 14.13 | 26.37 | 0.25 | 1.07 | 0.20 | 0.05 | 4.34 | 0.94 | 0.35 | 7.43 |
| LSV | 2 yellow | 56.08 | 15.54 | 15.90 | 0.19 | 0.76 | 0.14 | 0.06 | 3.56 | 0.83 | 0.17 | 6.78 |
| LSV | 3 yellow | 39.90 | 14.87 | 29.75 | 0.27 | 1.06 | 0.23 | 0.11 | 4.30 | 0.90 | 0.27 | 8.33 |
| LSV | 5 yellow | 44.94 | 15.77 | 24.17 | 0.82 | 0.95 | 0.29 | 0.10 | 3.80 | 0.80 | 0.18 | 8.17 |
| LP | 12 yellow | 59.51 | 13.65 | 14.71 | 0.16 | 1.03 | 0.20 | 0.09 | 4.11 | 0.82 | 0.30 | 5.41 |
| LP | 15 yellow | 60.31 | 18.74 | 7.33 | 0.04 | 1.25 | 0.22 | 0.09 | 6.10 | 1.13 | 0.23 | 4.56 |
| Milli | 27 yellow | 47.14 | 9.48 | 31.33 | 0.21 | 0.88 | 0.32 | 0.05 | 2.59 | 0.60 | 0.34 | 7.06 |
| Milli | 29 yellow | 55.06 | 16.43 | 15.66 | 0.12 | 1.50 | 0.42 | 0.04 | 5.50 | 1.06 | 0.29 | 3.92 |
| LSV | 4 green | 47.86 | 29.28 | 2.64 | 0.01 | 1.43 | 0.26 | 0.07 | 5.38 | 1.49 | 0.1 | 10.97 |
| LSV | 6 green | 57.73 | 22.38 | 3.22 | -0.01 | 1.53 | 0.30 | 0.10 | 7.05 | 1.53 | 0.10 | 6.07 |
| LSV | 8 green | 50.04 | 28.10 | 2.61 | 0.00 | 1.69 | 0.22 | 0.10 | 9.24 | 1.98 | 0.07 | 5.95 |
| Milli | 24 green | 66.74 | 17.60 | 4.89 | 0.00 | 1.39 | 0.07 | 0.04 | 7.17 | 1.21 | 0.03 | 0.86 |
| LSV | 5 red | 47.61 | 26.50 | 9.73 | 0.05 | 0.67 | 0.50 | 0.06 | 1.69 | 0.87 | 0.19 | 12.14 |
| LSV | 7 red | 39.08 | 24.96 | 17.81 | 0.23 | 0.71 | 0.58 | 0.05 | 1.47 | 0.61 | 0.18 | 14.31 |
| LSV | 9 red | 40.31 | 27.50 | 15.28 | 0.07 | 0.51 | 0.56 | 0.05 | 1.69 | 0.72 | 0.27 | 13.06 |
| LP | 20 red | 45.20 | 32.09 | 7.16 | 0.01 | 0.37 | 0.26 | 0.04 | 1.12 | 0.62 | 0.14 | 12.98 |
| Milli | 26 red | 44.82 | 29.34 | 7.27 | -0.02 | 0.86 | 0.68 | 0.08 | 2.05 | 0.74 | 0.44 | 13.71 |
| LP | 21 white | 46.57 | 34.52 | 1.52 | -0.01 | 0.83 | 0.17 | 0.04 | 3.64 | 1.37 | 0.05 | 11.29 |

Taken from Alló (2001)

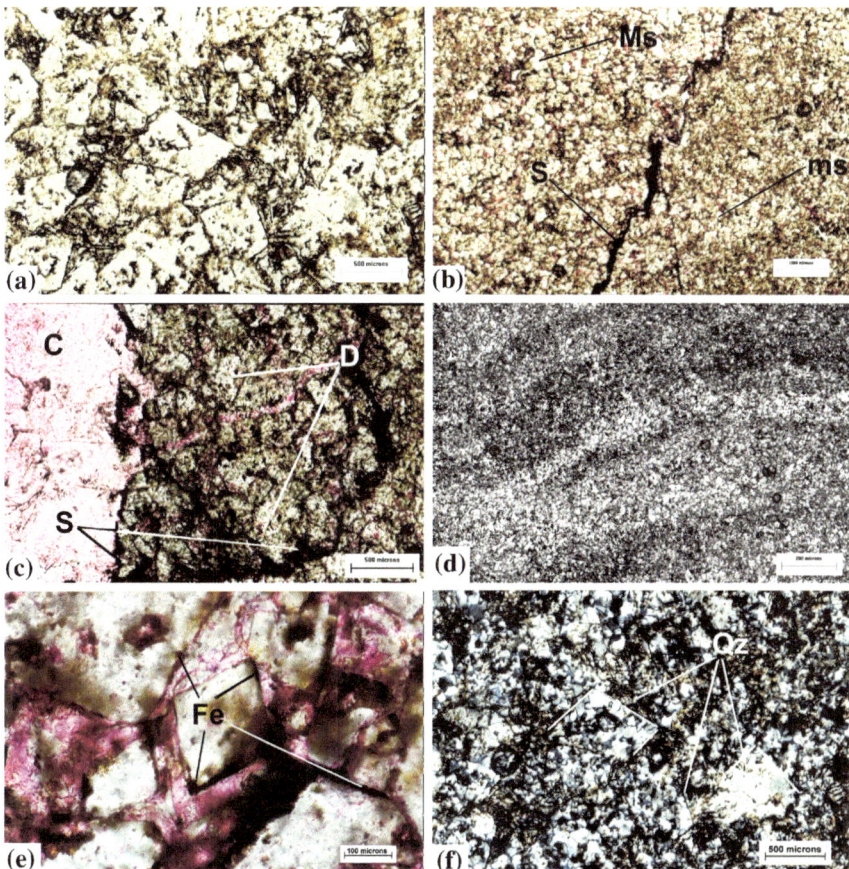

**Fig. 3.17** Photomicrographs of carbonate facies, Villa Monica Formation. La Placeres and Estancia la Siempre Verde quarries: **a** Dolostone stained with Alizarin red: bimodal, dolomitized rhombohedric carbonate crystals, on the base of La Juanita Formation. **b** Idem **a**: Ms: macrosparite. ms: microsparite, S: stylolite filled with ferric oxides. **c** Staining with Alizarin red shows that the dolomitization was complete. Sparry calcite only is present in fractures and in very small cavities. **d** Typical stromatolite lamination. **e** Calcite rims on dolomite crystals (*arrow*). Fe: ferric oxides. **f** Quartz pseudomorphous after dolomite rhombohedric crystals. Quartz: Qz. Taken from Zalba et al. (2010a)

rhombohedric dolomite crystals show silicified cellular structures which are interpreted as related to microbial activity. Quartz megacrystals (Fig. 3.18b) have grown in large cavities (fenestral porosity) and, in some cases, include opal-CT (Fig. 3.18c) identified as trydimite by X-ray diffraction (Table 3.12, sample 0). Cavities are large in comparison with the previous lithofacies and goethite is abundant, surrounding replaced dolomite crystals (Fig. 3.18b). When cavities increase in size they connect with each other and, occasionally, they are filled with quartz megacrystals. Typical cutans (Brewer 1960, 1976) are represented by

**Table 3.12** Quantitative X-ray diffraction analyses on total samples of La Siempre Verde, La Placeres and Don Camilo quarries (Rietveld method)

La Siempre Verde quarry

| Sample | Quartz | Kaolinite | Illite | Trydimite | Smectite | Goethite | Dolomite | Calcite | Rutile | Anatase |
|---|---|---|---|---|---|---|---|---|---|---|
| 6 | – | 81 | 19 | – | – | – | – | – | – | – |
| 5 | 4 | 40 | 36 | – | – | 17 | – | – | – | 3 |
| 4 | 28 | 10 | 15 | – | . | 47 | – | – | – | – |
| 3 | 19 | – | 1M$_1$; 15 2M$_1$; 61 | – | 2 | – | – | 1 | 3 | – |
| 2 | 24 | n.d. | 1M$_1$; 19 2M$_1$; 32 | – | – | 25 | – | – | – | – |
| 1 | 39 | n.d. | 41 | – | – | 20 | – | – | – | – |
| 0 | 10 | – | 24 | 46 | – | 20 | – | – | – | – |
| 00 | 5 | – | 5 | – | – | 3 | 79 | 8 | – | – |

La Placeres quarry

| Sample | Quartz | Kaolinite | Illite | Smectite | Goethite | Rutile | Anatase |
|---|---|---|---|---|---|---|---|
| 30 | 10 | 73 | 13 | – | 4 | – | – |
| 29 | 42 | 4 | 38 | – | 16 | – | – |
| 28 | – | 17 | 1M$_1$; 16 2M$_1$; 40 | 13 | 2 | 2 | – |
| 27 | 9 | 56 | 21 | 6 | 8 | – | – |
| 26 | – | 8 | 1M$_1$; 16 2M$_1$; 56 | 16 | 2 | 2 | – |
| 25 | 20 | 54 | 16 | – | 10 | – | – |
| 24 | – | 32 | 1M$_1$; 9 2M$_1$; 41 | 13 | 3 | – | 1 |
| 23 | 19 | 22 | 32 | 15 | 11 | – | 1 |

(continued)

**Table 3.12** (continued)

La Placeres quarry

| Sample | Quartz | Kaolinite | Illite | Smectite | Goethite | Rutile | Anatase |
|---|---|---|---|---|---|---|---|
| 22 | 37 | 8 | 36 | 6 | 13 | – | – |
| 21 | 24 | 12 | $1M_1$: 15<br>$2M_1$: 33 | 11 | 5 | – | – |
| 20 | 14 | 36 | 29 | – | 21 | – | – |
| 19 | 30 | 5 | $1M_1$: 20<br>$2M_1$: 31 | – | 14 | – | – |
| 18 | 55 | 2 | 23 | – | 20 | – | – |

Don Camilo quarry

| Sample | Quartz | Kaolinite | Illite | Smectite | Goethite | Rutile | Anatase |
|---|---|---|---|---|---|---|---|
| 14 | 37 | – | $1M_1$: 19<br>$2M_1$: 36 | – | 4 | 4 | – |
| 13 | 26 | 2 | $1M_1$: 27<br>$2M_1$: 39 | 2 | 20 | 4 | – |
| 12 | 18 | – | $1M_1$: 16<br>$2M_1$: 27 | – | 30 | 1 | – |
| 11 | 17 | 7 | $1M_1$: 8<br>$2M_1$: 37 | 3 | 27 | 1 | – |
| 10 | 49 | – | 24 | – | 27 | – | – |
| 9 | 38 | 1 | $1M_1$: 26<br>$2M_1$: 34 | – | -3 | 1 | – |
| 8 | – | – | – | – | – | – | – |
| 7 | 65 | 4 | 10 | 2 | 19 | – | – |
| 6 | 16 | – | 10 | – | – | – | – |

(continued)

**Table 3.12** (continued)

Don Camilo quarry

| Sample | Quartz | Kaolinite | Illite | Smectite | Goethite | Rutile | Anatase |
|---|---|---|---|---|---|---|---|
| 5 | 29 | 42 | 10 | 2 | 17 | – | – |
| 4 | 7 | 2 | 1M$_I$: 33<br>2M$_I$: 55 | – | – | 3 | – |
| 3 | 6 | 9 | 7 | 13 | 78 | – | – |
| 2 | – | 3 | 1M$_I$: 26<br>2M$_I$: 68 | – | – | – | – |
| 1 | 26 | tr | 37 | – | 34 | 1 | 1 |

Taken from Zalba et al. (2010a)

ferriargillans and clay infillings. Ferriargillans show spherical or ellipsoidal shapes, parallel lamination and are rich in ferric hydroxides (goethite), kaolinite, and minor smectite showing orange to yellowish colors. Orientation of the clay laminae produces extinction when parallel to the polarizers (Fig. 3.18d). Ferriargillans fill cavities developed among relics of the original rhombohedric dolomite crystals, now replaced by illitic material and disrupt, black, hematized microbial mat deposits (MM: Fig. 3.18d). Two types of clay infillings, which block up cavities, were distinguished: (1) brownish kaolinite (Table 3.12, sample 20), lacking rutile needles and showing a book texture. They are discernible by their interference color from (2) illitic clay (dark gray) which is composed of a mixture of $1M_1$ and $2M_1$ polytypes (Table 3.12), where $2M_1$ predominates over $1M_1$.

Kaolinite, associated with smectite and goethite, fills fractures and microfractures with different orientations as has been already shown in the field (Fig. 3.15b) cutting the weathered dolostones at all levels and showing slickensides (stress cutans). The ferriargillans and the original dolomite texture are also cut by microfractures filled with kaolinite, smectite, and goethite. Detrital quartz grains, with overgrowths and undulate extinction, show some orientation within the relic dolostones and also seem to form defined siliciclastic intercalations. In Fig. 3.18e, crinkled dark microbial mat deposits (MM) alternate with illitized carbonate deposits (light color). Dessication cracks in kaolinitic, smectitic clays (white) are clearly observed in microfractures cutting microbial mats (in dark).

In Zalba et al. (2010a), the authors stated that the siliciclastics in the C/S facies are represented by silt to clay-sized layers and by minor sand deposits. The fines may show lamination, graded structure, and stylolites parallel and inclined to bedding, either filled with clay, sand and/or hematite, the perpendicular stylolites cutting the parallel ones Fig. 3.19a). Also, the quartz sandstones show great compaction, sutured boundaries, and concave–convex contacts between grains, undulate extinction, fluid inclusions, secondary silica growths, and dissolution features (Fig. 3.19b). Deformational features (e.g., folding) are common in these rocks, which show connected cavities filled either with clays or quartz cement. Quartz grains show dissolution effects. Some quartz crystals have also been broken and pulled apart by the introduction of clay, as shown along stylolite surface (Fig. 3.19c). The clay introduced between grains is kaolinite.

As shown by X-ray diffraction, the same authors sustain that silt to clay-sized beds are composed of predominantly illitic material (I + ISII) with <15 % of expanded layers. Quantification of $1M_1$ and $2M_1$ illite polytypes shows that the $2M_1$ polytype predominates over $1M_1$ in the total sample (see Table 3.12). Illitic material bears rutile needles (5–30 µm long) arranged in a sagenite-like texture (Fig. 3.20). Opal-A, considered of biogenic origin, recrystallized to opal-CT (trydimite), then to quartz megacrystals (which include opal-CT) which developed in fenestral cavities. After the silicification process, subaerial exposure occurred. The sediments must have been eroded to a large extent and extensive weathering occurred with the development of pedogenetic features (e.g., clay illuviation). Dissolution of the boundstones increased and cavities enlarged and connected with each other. Loose, individual or random aggregates of pyramidal quartz

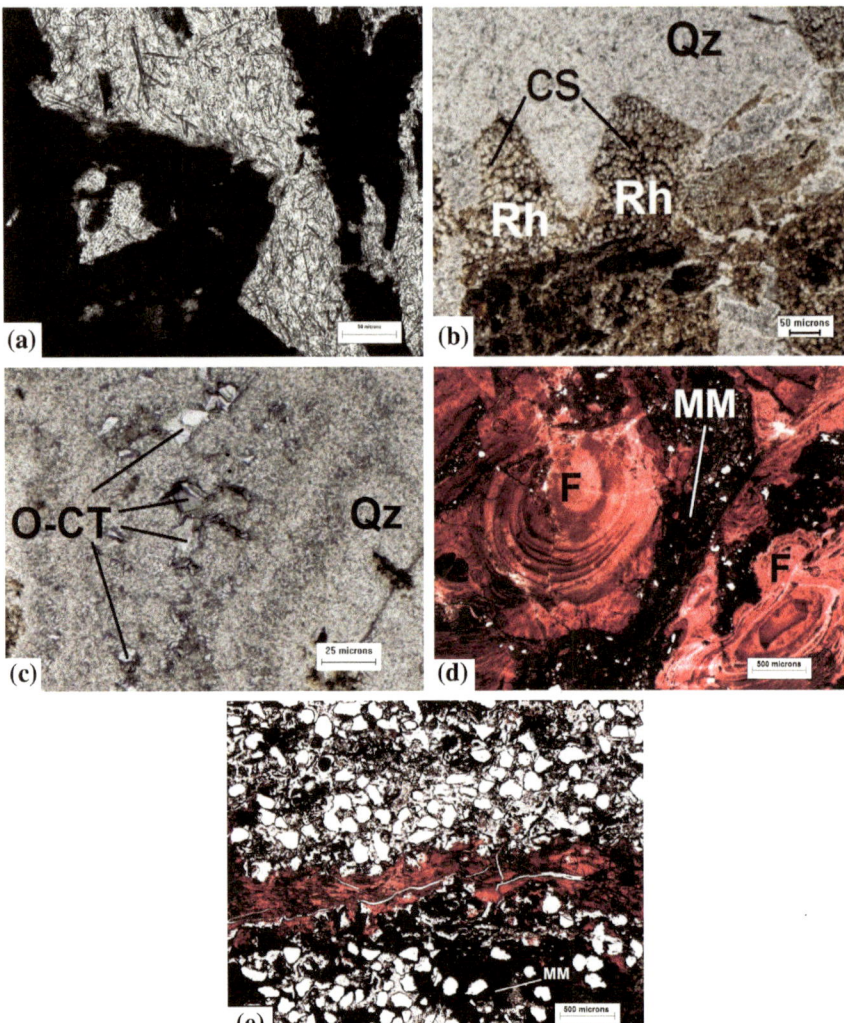

**Fig. 3.18** Photomicrographs of carbonate-siliciclastic facies, Villa Monica formation. La Placeres and Estancia la Siempre Verde quarries: **a** Illitic clays with rutile needles in a sagenitic-like texture, filling cavities and replacing relic dolomite rhomboedric crystals. **b** Quartz megacrystals (Qz) include rhombohedric dolomite crystals (Rh) which show cellular structure (cs). **c** Detail. Quartz (Qz) includes opal-CT (O-CT). **d** Ferriargillans (F), some of them with spherical shapes, showing parallel lamination and filling cavities developed among relic microbial mats (in *black*: MM) with isolated, detrital quartz grains. **e** Fractures filled with ferriargillans (middle part) crossed by smaller fractures filled with kaolinite. Lower part (in *black*: MM): relic of microbial mat deposits with detrital quartz grains trapped within. Upper part: siliciclastic deposit with abundant detrital quartz grains floating in an illitic epimatrix. Taken from Zalba et al. (2010a)

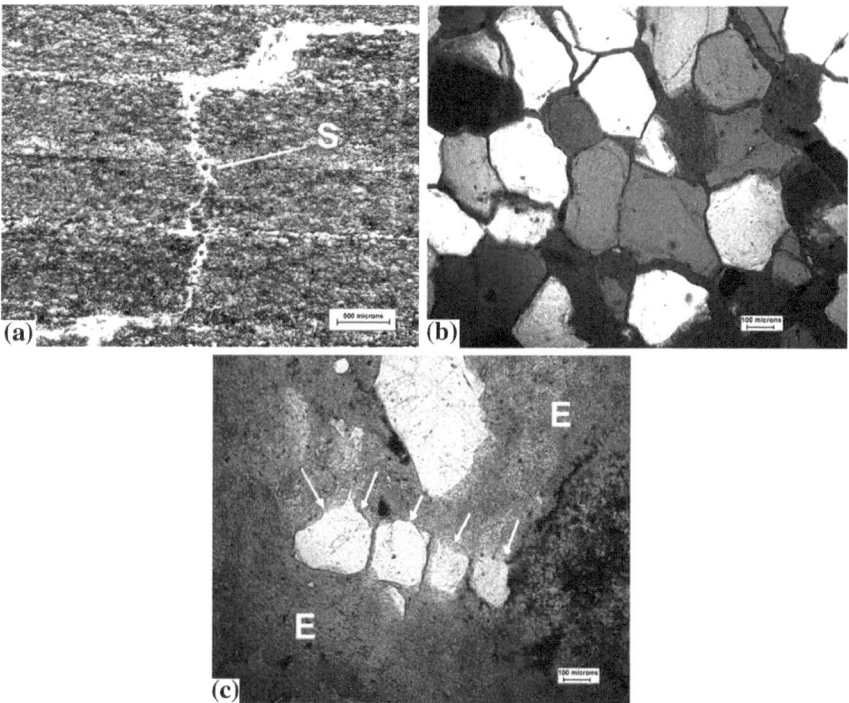

**Fig. 3.19** Photomicrographs of carbonate-siliciclastic facies, Villa Monica Formation. La Placeres and Estancia la Siempre Verde quarries. **a** Stylolite (S) perpendicular to bedding cut graded illitic clay levels. Stylolite is filled with detrital material. **b** Quartzite. Detrital quartz grains, with secondary overgrowths, have been broken and pulled apart by the infiltration of clays (*dark rims*). **c** Dissolutions effects (arrows) due to the development of a stylolite surface on detrital quartz grains, also disrupted by the introduction of epimatrix (E). Taken from Zalba et al. (2010a)

megacrystals remained "in situ" in the weathered dolostones (C/S facies). Illuviation processes deposited cutans (illitic clay skins: argillans) in the created cavities. As the original rhombohedral dolomite crystals were dissolved, illuviated illitic clays increased in lower horizons. The ferric oxides may have derived from the dissolution of micas (biotite) contained in higher eluviated horizons, from the replacement of calcium by iron in the dolostones (Larsen and Chilingar 1979) and from the activity of cyanobacteria (blue-green algae) cohabiting on microbial film or mat (Dai et al. 2004). Detrital clay beds within the weathered dolostones (C/S facies) are represented by major $2M_1$ illite and subordinate $1M_1$ illite, the latter considered to be of diagenetic origin. Some of the clays show no quartz and a face-to-face association of platelets, which were deposited as flocs (card-pack-flocs, Williamson 1980). They show "swirl patterns", probably resulting from movements during packing, settling, or soft sediment deformation (Keller 1978).

The fines of the overlying heterolithic (H) facies are illitic and bear quartz, which increases upwards. The same illite polytypes are present and considered to be of the

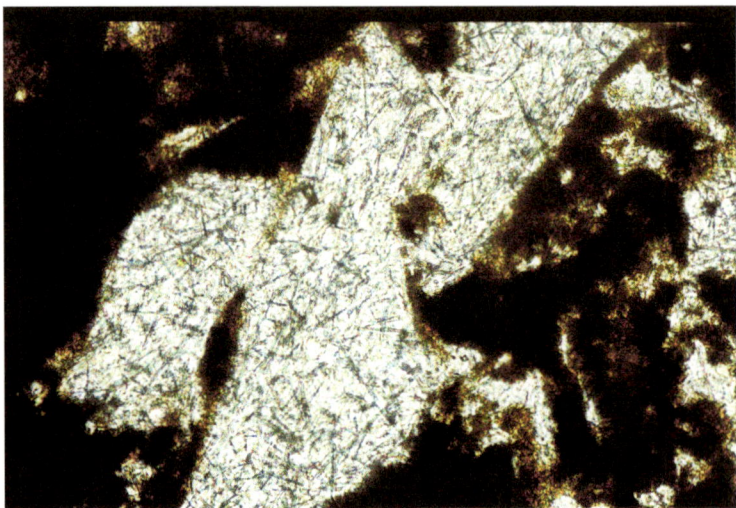

**Fig. 3.20** Photomicrograph of carbonate-siliciclastic facies. La Siempre Verde quarry. Illitic clay bearing rutile needles in a sagenitic-like texture. Taken from Zalba et al. (2010a)

same origin as in the previous illitic clays: $1M_1$. The upwards increasing trend in quartz content in the (H) facies is consistent with littoral conditions in tidal flats with the development of graded structures due to differential settling from suspension after a flood (Manassero et al. 2012). During burial diagenetic stages, diverse processes took place in the different lithofacies. Neoformation of minerals is represented by silica overgrowths and the development of (ISII) with <15 % expansive layers and I/S ($1M_1$ illite) at the expense of detrital feldspars (perpendicular clay coatings). It has to be taken into account that ISII derive from the alteration of $2M_1$ illite and muscovite (detrital). This $2M_1$ illite is never found as a diagenetic product. When illite derives from feldspar or other minerals, the polytype is $M_1$ and it is diagenetic in origin (Pevear 1999). Other neoformed minerals are rutile, pyrite, and minor anatase. Rutile precipitated from titanium oxide contained in the micas, as in many other Neoproterozoic formations of Tandilia (Iñíguez et al. 1989) and crystallized as needles (resembling a sagenite-like texture) in the illitic clays and in quartz cement, filling cavities or fractures. Rutile needles are also found in quartz precipitated in pores of the quartzites of the (H) facies. Pressure-solution by overburden originated stylolites parallel to bedding and intergranular (e.g., concave–convex grain contacts) dissolution effects developed in the C/S as well as in the H facies.

It is important to underline that kaolinite has been introduced in the sediments because it is associated with minor smectite in ferriargillans and clay infillings related to a crisscrossing network system of fractures cutting the C/S facies and postdating the illuviation of illitic clay infillings (argillans). It represents the redistribution of neoformed clays in the sediments and in the sedimentary discontinuity between the C/S and the H facies. This would explain the mixture of the kaolinite-halloysite assemblage and/or smectite, with illitic clays which, almost certainly, is the same phenomenon that occurred in most studied clay deposits of Tandilia where this mixture of illite with kaolinite-halloysite is found (Zalba et al. 2007a).

Smectite, in variable proportions, is predominately associated with slickensides in reddish clays filling a network system of fractures. The slickensides differ markedly from clay skins, which occur on pedogenetic surfaces resulting from clay translocation. The latter have sharp outer and inner boundaries with distinct extinction patterns and are often finely layered (laminar fabric). The slickensides seem to be aligned preferentially (relative to the stress field) in response to structural deformation (Driese and Foreman 1992), so they would be of tectonic origin instead of being pedogenetic, which are not oriented and locally form pseudo-anticlines (Driese and Foreman 1992; Driese et al. 1992; Driese and Mora 1993; Caudill et al. 1996). Smectite may derive from the dolostones which contain the Mg and Fe required for its development as a result of neogenesis or transformation from primary minerals (Worden and Burley 2003). If smectite had formed earlier than the other clay minerals (as in vertisols) it would have prevented the illuviation (translocation) of illitic clay throughout the weathered boundstones. Furthermore, the presence of smectite-kaolinite-halloysite filling fractures suggests a late origin for these minerals. Smectite was one of the latest minerals to form and must have developed during the telogenetic stage. This mineral is also present in other Neoproterozoic units of the Tandilia Basin (e.g., Cerro Negro Formation) and was considered to be of late diagenetic origin (Zalba 1982).

Cubic crystals of pseudomorphous hematite after pyrite could have formed during this telogenetic stage due to oxidizing meteoric fluids. Pyrite was presumably formed during early burial diagenesis, as in other Neoproterozoic deposits of Tandilia (Zalba et al. 2007a). Hematite which colored the sediments and is also present as diffusion cutans needs oxidizing conditions to be formed. Most of the hematite was hydrated to goethite, the latter being a constituent of the ferriargillans. It is present in fractures, in cavities of the weathered boundstones and in the siliciclastics of the C/S facies. Fractures filled with goethite cut the remains of the original boundstones, the ferriargillans, and the clay skins in general, demonstrating that goethite post-dated all of them. Petrography shows that goethite grew over smectite and so the first must be younger.

Lateral compressive forces produced pressure-solution effects. Inclined stylolites developed in all the identified facies cutting previous stylolites developed parallel to bedding. They are filled with epimatrix and/or relic ferric oxides (hematite), dissolved and fractured quartz grains and brecciated ferriargillans, which indicate compressive effects.

Folding of the sediments also occurred during telogenesis and affected in a different way the lithostratigraphic units because of a different mechanical response of these materials to compressive forces. This folding is clearly appreciated in the field and through microscopic observations. Another event attributed to telogenesis is the development of kaolinitic epimatrix (tangential clay coatings), clearly connected with fractures, surrounding secondary quartz overgrowths and observed in sandstones of the H facies.

Dedolomitization partially affected the C facies. According to Larsen and Chilingar (1979), dedolomitization can be eogenetic as well as telogenetic. In the present case, dedolomitization is recognized by the presence of epitaxial calcite rims on silicified dolomite crystals in the C facies. The rims are even and uniform in thickness. Analysis of the fabric geometry and mineral paragenesis suggests that the rims formed by marginal dedolomitization may be attributed to telogenetic processes.

Calcification (passive precipitation) also occurred during this telogenetic stage. Calcite fills fractures enlarging them because of its force of crystallization. These fractures cut stylolites, thus postdating them. Calcite also fills large cavities in trapped sediments in the C and C/S facies.

### 3.3.3  Microscale Diagnostic Diagenetic Features in the Villa Mónica Formation

In *Microscale diagnostic diagenetic features in Neoproterozoic and Ordovician Units, Tandilia Basin, Argentina: A Review*, Zalba et al. (2010b) made a summary of diagnostic diagenetic features and veiled biosignatures found in siliciclastic and mixed facies of the Villa Mónica Formation, La Juanita Sector which will be discussed in this chapter. Textural studies were carried out on clays and in well preserved and weathered dolostones (Table 3.13).

As sustained by Gerdes et al. (1991) "microbial mat formation on bedding planes may support the separation of sedimentary units". In the case of the Villa Mónica Formation, for example, rare microbial structures found some years ago at San Manuel area (see map of Tandilia) were only understood after having compared them with other more complete and preserved recently found in other localities of the Tandilia System. When some structures in the mixed facies (C/S) of the Villa Mónica Formation were interpreted to be biogenerated (Zalba et al. 2010a; Manassero et al. 2012) new petrographical observations in the Neoproterozoic Villa Mónica Formation were carried out searching for biosignatures. The authors followed the classification of the diagenetic processes through the classical scheme of early diagenesis (eogenesis), burial diagenesis (mesogenesis), and uplift-related diagenesis (telogenesis) according to the scheme developed at first for limestone diagenesis by Choquette and Pray (1970), and later generalized for clastic diagenesis. Weathering processes involved were also taken into account.

According to Zalba et al. (2007a) the sedimentary sequences of Tandilia have probably not exceeded 2–3 km of burial. This statement is based on the comparatively simple authigenic mineralogy of these successions (cf. Iñiguez and Zalba 1974; Iñiguez et al. 1989; Zalba 1982), also revealed on several world examples of onshore basin-margin sequences (Hebridean Basins, Morton 1987; Yorkshire Basin, Hemingway and Riddler 1982; Dorset Basin, Scotchman 1991a, b), which, according to these authors, experienced a 2.5 km of burial. Furthermore, the

**Table 3.13**  Stratigraphic units and sedimentary cycles of the Tandilia Basin

| Age | Sierras Bayas | | Barker | | Sedimentary cycles | |
|---|---|---|---|---|---|---|
| | Area | | | | | |
| Early Ordovician | | | Balcarce Fm. (quartzite) | | Batán ∼∼∼∼ | δ |
| Neo-proterozoic | Cerro Negro Fm. (claystone, marl, and limestone) | | Las Aguilas Fm. (breccia, claystone, and quartzite) | | La Providencia ∼∼∼ | δ |
| | Sierras Bayas Group | Loma Negra Fm. (limestone) | Sierras Bayas Group | Loma Negra Fm. (limestone) | Villa Fortabat ∼∼∼ | δ |
| | | Olavarría Fm. (claystone) | | Las Aguilas Fm (?) (breccia, claystone, and quartzite) | Diamante ∼∼∼ | δ |
| | | Cerro Largo Fm. (quartzite) | | Cerro Largo Fm. (claystone, and quartzite) | Malegni ∼∼∼ | δ |
| | | Villa Mónica Fm. (dolostone, claystone, and quartzite) | | Villa Mónica Fm. (dolostone, claystone, and quartzite) | Tofoletti ∼∼∼ | δ |
| Paleo-proterozoic | Complejo Buenos Aires (granitoids) | | | | ∼∼∼∼ | δ |

Taken from Iñíguez et al. (1989), modified by Andreis et al. (1996) and Poiré and Spalletti (2005)
(?): Indicates the location for the Las Aguilas Formation proposed by Poiré and Spalletti (2005) who correlates this unit with the Olavarría Formation

abundance of compaction features like straight-line and suture boundaries and especially the presence of triple junctions in sandstones (Ahmad and Bhat 2006) observed in the quartzites of the Sierras Bayas Group (and of the Balcarce Formation: Iñíguez et al. 1996) support this assumption.

At present there are still many geological problems to solve but, according to recent studies (Zalba et al. 2007b, 2010a, b; Manassero et al. 2012), detailed petrographical work and the interpretation of paragenetic sequences, presented next, have proved to be of the most importance in the better understanding of many processes which have occurred in these ancient rocks. Cellular microbial colonies and crinkled, banded lamination (dark) can be seen in dolomite rhombohedra (in brown) where cavities have been filled with quartz (see Fig. 3.18b). Zalba et al. (2010b) interpreted that cellular/banded stromatolitic structures and crinkled lamination are the kind of deposits derived from bacteria and blue-green algae activity, according to Schieber (1998). Cathodoluminiscence analysis performed on the previous sample (Fig. 3.21a) shows dolomite with microbial colonies (in orange) included in quartz crystals and in Fig. 3.21b, quartz crystal grown in dolomite cavities. The quartz crystals include dolomite rhombohedra, again suggesting that

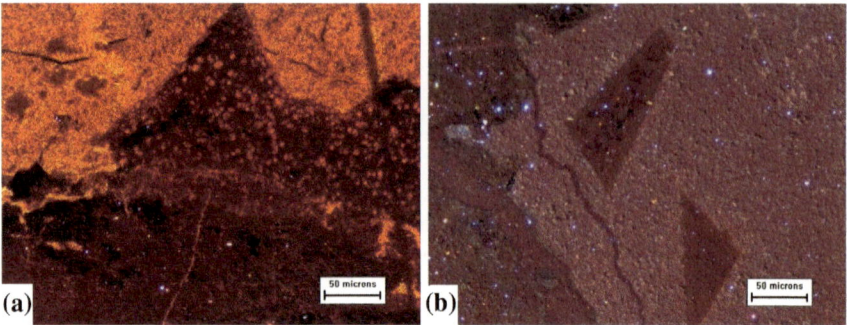

**Fig. 3.21** Cathodoluminescence images. Villa Mónica Formation, La Siempre Verde quarry.
**a** Dolomite with microbial colonies (in *orange*) included in quartz crystals (in *brown*). **b** Quartz
crystal grown in cavities of stromatolite dolostones which include dolomite rhombohedra (in
*brown*). Taken from Zalba et al. (2010b)

dolomitization preceded silicification processes. In conclusion, calcite precipitation
from cyanobacteria, precipitation of opal-A, opal-CT (trydimite), quartz (silicification), and dolomitization are considered textural diagnostic characteristics of
syngenesis and eogenesis.

A relatively new concept linked to structures found in bioconstructed rocks is the
recognition of Microbially Induced Sedimentary Structures: MISS (Noffke et al.
2001), which, according to the original definition of the authors "do not arise from
chemical processes, but from the biotic-physical interaction of microbial mats with
the sedimentary dynamics of aquatic environments". New defined mixed
(carbonate-siliciclastic) facies for the Neoproterozoic Villa Mónica Formation show
evidence of biosignatures (Zalba et al. 2010a) which were described later (Zalba
et al. 2010b) as MISS.

Microbial mats are well known from stromatolites in carbonates back to 3.5 Ga
(Walter 1994). Many contributions on stromatolites in Proterozoic basins are
applied to carbonate environments (Awramik 1984; Walter et al. 1992) while there
are few references to clastic environments (Schieber 1998; Noffke et al. 2003;
Noffke 2006, 2009). Siliciclastic stromatolites are scarce compared to their carbonate counterparts in the rock record. The paucity of stromatolites in siliciclastic
strata may relate to physical conditions and processes within the depositional
environment that inhibit stromatolite growths and preservation (Druschke et al.
2009).

MISS (Fig. 3.22a) are represented by thin sedimentary units (from microns to
few millimeters) composed of two biofabrics: (1): a characteristic dark, ferric,
wavy, crinckly microbial mat deposit, with silt to clay-sized detrital grains trapped
within (Noffke et al. 1997), interlayered with (2): a micritic, originally carbonate
layer (light part of the photograph) with relics of rhomboedral illitic clay subsequent
to dolomite enclosed by hematite and where bent micas, signifying detrital origin,
are observed. On the one hand, no biogenic structures have remained after

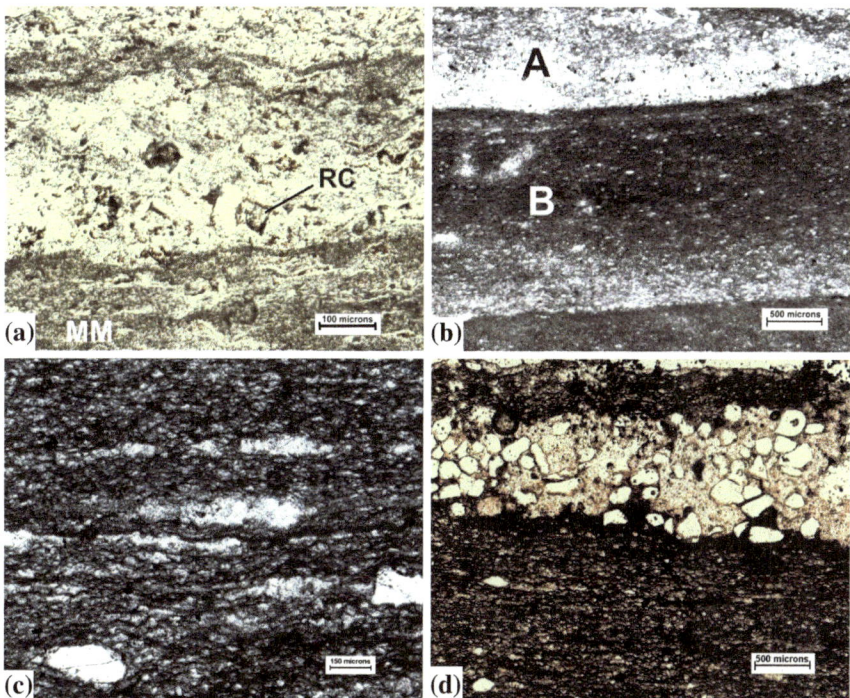

**Fig. 3.22** Photomicrographs of MISS in Carbonate/Siliciclastic facies. Villa Mónica Formation, Don Camilo quarry: **a** Dark, hematized, wavy crinkly microbial mat laminae (MM) alternating with light parts containing illitized dolomite with ghost rhombohedral crystals (RC) (PPL). **b** Part (A) Basal, grain-supported silt to sand-sized sediments with graded structure and minor interlayered microbial mats. Part (B) Dark, wavy, crinkly microbial laminae with detrital grains 'floating' within and all oriented with their long axis parallel to the bedding plane (XPL). **c** Dark, ferric, crinkly microbial mat laminae with isolated detrital grains within the mats oriented parallel to bedding plane (XPL). **d** Quartz grains 'float' in an illitic epimatrix and alternate with microbial mats (*dark* part). Detrital quartz grains oriented with their long axis parallel to bedding planes (PPL) Zalba et al. (2010b)

carbonate neogenesis in the light part of the unit, presumably because of recrystallization. On the other hand, microbial mat preservation (dark part of the photograph) has been possible due to the precipitation of ferric oxides which did not allow the expansion of diagenetic carbonates as suggested by Hofmann (1975).

In the siliciclastics of the Carbonate-Siliciclastic facies of the Villa Mónica Formation we can differentiate bioconstructed structures from purely physically deposited siliciclastic beds. In Fig. 3.22b two different parts can be recognized: Part (A) Basal, grain-supported silt to sand-sized sediments with graded structure and minor interlayered microbial mats, showing some grains orientated preferentially with their long axis parallel to the sedimentary surface. Part (B) Upwards, the siliciclastic material decreases and a dark, curvy, crinckly lamination with detrital material floating within becomes more profuse. These dark deposits represent microbial laminites with great pigmentation (ferric oxides). Within the microbial mats, all the detrital quartz grains are oriented with their long axis parallel to the sedimentary surface. In this framework, these structures are considered part of the MISS, as defined by Noffke et al. (2001). The graded structure could be preserved because the microbial laminae have acted as a paste, occupying, trapping, stabilizing the sedimentary surface and preserving it, according to Noffke et al. (1997). In the upper intertidal and lower subtidal zone of tidal flats, microbial mats stabilized the sedimentary surface (Krumbein 1994). The features of microbial mats shown on vertical thin sections (dark colored, intact and wavy, wrinkled laminae) are important indicative characteristics to distinguish similar wrinkle structures but originated by abiogenic processes, as stated by Noffke (2007). While the laminar morphology of the iron-rich laminites may not preclude an abiologic origin, the concentration of elements such as iron in sedimentary laminae is commonly attributed to microbial metabolism (Flugel 2004). Figure 3.22c is another example of MISS where dark, ferric, crinkly microbial mat laminites with isolated, detrital grains within the mats oriented parallel to bedding plane are illustrated. Alternating, graded, well-sorted and well-rounded sandy to silty siliciclastic deposits, with some orientation of their long axis parallel to the bedding plane are shown. In Fig. 3.22d quartz grains float in an illitic epimatrix and alternate with microbial mats (dark part) where detrital quartz grains oriented with their long axis parallel to bedding planes are clearly observed. Microbial mat formation in bedding planes may support the division of sedimentary units, an important concept supported by Gerdes et al. (1991) and which was taken into account for the detection and understanding of microbial mat relics in the named "ferruginous clays" (Manassero 1986), also known as Psamopelites (Poiré and Iñiguez 1984) of the Villa Mónica Formation at La Placeres. That is why based on field work and petrological studies these rocks were reclassified as mixed facies (Zalba et al. 2010a).

MISS are singenetically formed and they have been preserved through silicification and dolomitization (eogenesis); mesogenesis (e.g., neoformation and transformation of minerals, compaction) and also through telogenesis. The presence of iron oxides and dolomitic cements suggests microbial mat mineralization. In modern environments photosynthetic cyanobacteria may produce oxygen at the surface of microbial mats but, directly below the oxic surface layer, anaerobic bacteria degrade mat generated organic matter and may create a localized, strongly reducing environment (Gerdes et al. 2000). These environmental conditions favor the precipitation of calcium and ferroan carbonates and pyrite (Schieber 2007; Schieber and Riciputi 2004). Such minerals, and also oxidized forms of pyrite (hematite and goethite) can be a valuable indicator of the former presence of microbial mats (Noffke et al. 2006; Schieber 2007). In the case of the examples displayed in Fig. 3.23a, b, hematization during early burial has enabled the preservation of discontinuous microbial mat relics preserved as lonely witnesses of original bedding planes.

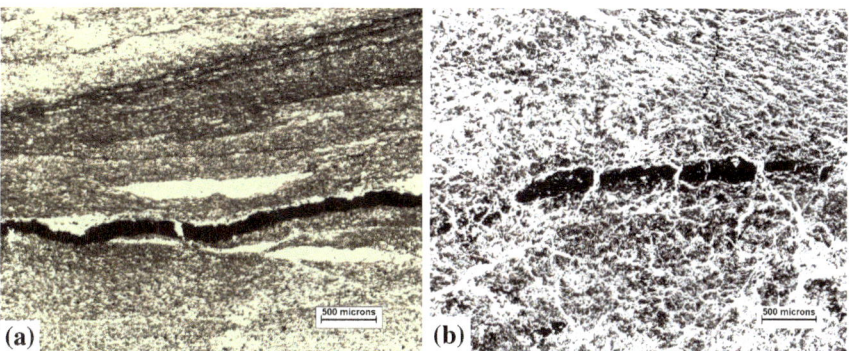

Fig. 3.23 Photomicrographs of relics of MISS. **a** Dark, hematized microbial mat relics in mixed facies (Carbonate/Siliciclastic). Villa Mónica Formation, La Placeres quarry, Sierra La Juanita, Barker. **b** Fractured MISS where epimatrix has infilled cracks in well-preserved dolostones, Villa Mónica Formation, La Siempre Verde, Sierra La Juanita, Barker. (PPL). Taken from Zalba et al. (2010b)

### 3.3.4 Scanning Electron Microscopy of Clays of the Villa Mónica Formation

SEM shows the texture of illitic green clays, intercalated in fresh dolostones as well as in weathered dolostones of the carbonate-siliciclastic facies; with a face-to-face orientation of the laminae, typical design of marine deposition (Fig. 3.24a).

Accordions of diagenetic, red kaolinite clays formed in cracks and fissures and in available pore space in the carbonate-siliciclastic facies (Fig. 3.24b). Diagenetic goethite forms rounded aggregates of laminar crystals in weathered dolostones (Fig. 3.24c) and diagenetic tubular halloysite transformed from previous kaolinite, both formed in fissures which crisscross the carbonate-siliciclastic facies (Fig. 3.24d). Diagenetic kaolinite transforming into tubular halloysite, and goethite, are interpreted as formed in a telogenetic stage.

### 3.3.5 Paleoenvironmental Conditions of the Villa Mónica Formation at Sierra La Juanita

Manassero et al. (2012) proposed—together with two of the authors of this piece of work—an interpretation of the paleoenvironmental conditions which could have occurred during the deposition of the Villa Mónica Formation in the studied area of the Sierra la Juanita and which will be exposed in the next pages. The carbonate facies located to the base of this unit include five meters of dolopackstone/wackestone described previously in the La Siempre Verde Section (Fig. 3.13). This carbonate platform composed of abundant stromatolites of

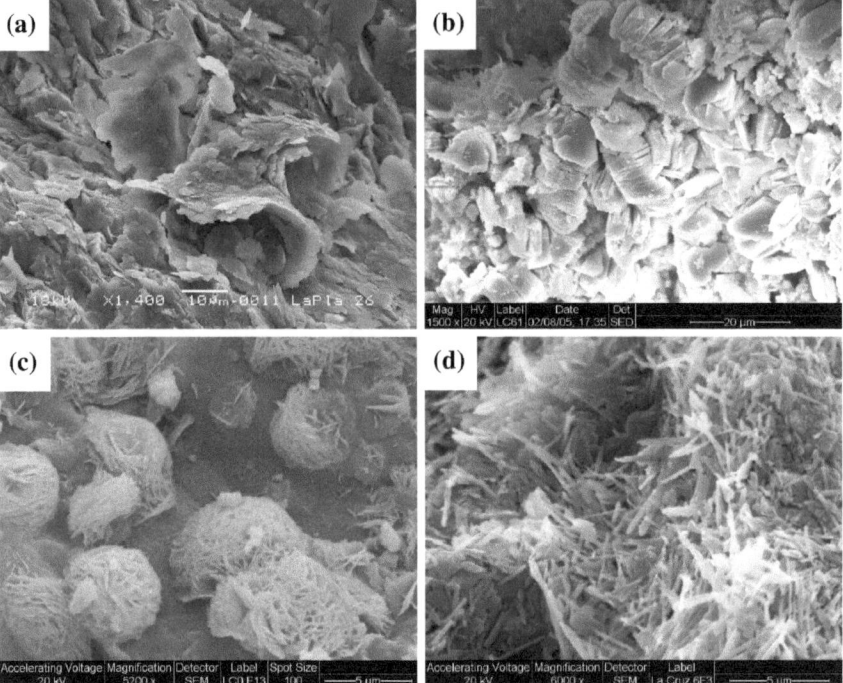

**Fig. 3.24** Scanning electron micrographs of carbonate-siliciclastic facies of Villa Monica Formation. **a** Texture of illitic green clays with stacked sheets face-to-face, by deposition in the marine environment. **b** Accordions of diagenetic, red kaolinite clays formed in cracks, fractures and filling pore spaces. **c** Diagenetic goethite in masses of laminar crystals on weathered dolostones. **d** Diagenetic halloysite, growing into tubes from previous kaolinite crystals, both formed in fissures. The *bar* indicates the scale photo. Micrographs 3.24 a was taken from Zalba et al. (2010a)

dominant columnar and domal shapes and sizes up to 30 cm high already shown (Figs. 3.14a and 3.15a) similar to the ones described by Jiang et al. (2003) for a tidal Proterozoic carbonate platform within a passive margin setting in Asia. The columnar stromatolites are commonly linked with peloids, ooids, and intraclasts present in lows between stromatolite heads. The relatively high relief limits interference and generates these simple shapes like columns and domes seen in the study area. On the other side, low relief permits sediment to interfere with accretion and generates branching (Riding 2011). These facies are interpreted to have accumulated in a relatively undisturbed shallow subtidal to lower intertidal environment.

In the carbonate-siliciclastic deposits the grain-supported intervals of the siliciclastic/microbial mat units indicate the initial deposition of coarse grains from suspension in a relatively upper flow regime. When flow velocity decreases the grain size decreases gradually to finer sediments. The origin of grading in intertidal flats has been attributed to deposition in the last phase of heavy floods (Reineck and Singh 1986). As stated by Noffke et al. (1997), and also sustained by Kah et al.

(2009) in modern environments the orientation of the grains within the mats with their long axis parallel to the bedding plane points out to an energetically suitable position following gravity forces, made possibly by the friction reduction of the soft organic matter. According to these authors the particles in the grain-supported part of the sedimentary unit (siliciclastic-microbial mats) do not show predominant orientation of their long axis parallel to the bedding plane since the compact fabric may not allow for their arrangement according to gravity. We agree with Noffke et al. (1997) in that "build-up" of these types of successive sedimentary units is a response to alternate depositional and non-or-low-rate depositional events. No reworking of the surface of each unit is perceived, since microbial mats prevent erosion during periods of increased flows. Otherwise, the units would lose their appearance and would have been amalgamated. A specific character of the occurrence of microbial mats is the arrangement of detrital quartz grains within the organic layers and the presence of separation surfaces between sedimentary units, whose preservation is precisely attributed to microbial activity and which result in bedding planes in consolidated rocks after burial. Many authors coincide in that the recognition of microbial mats in ancient terrigenous sediments is a difficult task, due to the fact that mat morphology is obliterated first by depositional processes or bioturbation and later by compactional processes (Schieber 1998). The interpretation of fine-grained Precambrian stromatolites is a challenge, as we do not know if they are agglutinated or precipitated and which microbes are involved, and also many morphotypes of these Proterozoic bodies have no modern analogues. They occur in a variety of modern tropical environments, from humid shores where they pass laterally towards freshwater marsh or to evaporite sabkhas (Riding 2000).

The potential of preservation is also controlled more by the presence of these microbial communities than by physical factors, because sediment surfaces colonized by microbes are less erodable. Features like wrinkle structures (see Fig. 3.16b) as well as erosional marks and microbial sand chips are formed by tractional currents in intertidal and supratidal zones (Bouougri and Porada 2002).

Considering the balance between the proportions of sediments and the development of microbial laminae it is possible to infer that the detrital short-lived continental input was not strong enough to eliminate the microbial colonies, but allowed them to grow in thin cycles. The fact that concavities and convexities of microbial mats do not superpose upwards in the sequence would mean that the growth of the microbial mats would have to be interrupted by external periodically controlled events (environmental) occurring in tidal flats such as seasonal episodes (Hofmann 1975). When buried by sand the bacteria quickly migrated upward towards the new sedimentary surface, where they established new mat fabrics (Noffke 2007). The microbial mats usually develop under translucent quartz that conducts light into deeper portions of the biofilms in sites with moderate hydraulic reworking. These biofilms must be able to tolerate the physical sediment dynamics caused by waves and currents (Noffke 2009). The recognition of a mixed origin for the carbonate-siliciclastic and heterolithic lithofacies by Zalba et al. (2010a, b), considering a microbial control versus pure detrital processes in the sedimentary record was decisive in the right assessment of the paleoenvironmental conditions and also confirmed the rightness of considering the heterolithic facies as part of the Villa Mónica Formation, instead of previous interpretations (Poiré and Iñiguez 1984) which assigned these sediments to the overlying Cerro Largo Formation. They contain abundant traction structures, such as symmetrical wave ripples suggesting deposition up to the breaker zone.

The association of mud-silt-sand tidal deposits with flaser and lenticular bedding with wrinkle and MISS has been studied in detail by Noffke (2009).

The traditional model of peritidal carbonate sedimentation on continental shelves and epeiric seas (Iñiguez et al. 1989; Cingolani 2011), regardless of age, is a shoreline model based on modern analogues of coastal tidal flats in various tropical areas. According to this concept, peritidal sediments are considered to fringe the land surface or the lee sides of reefs or grainstone shoals, or as in this case large, shallow epeiric seas (Andreis et al. 1996) form laterally continuous regionally broad belts many tens or even hundreds of kilometers in width (Pratt and James 1986).

All the lithofacies described are interpreted then as representing a prograding carbonate sequence dominated by tidal processes. The sea was opened to the west during Precambrian times and the coast line had an N–S trend which is coherent with regional sandstone paleocurrents data provided by previous authors (Andreis 2003). It is important to underline that due to the pericratonic location of these epeiric seas described above, these peritidal carbonates could have been exposed geographically and experimented important facies changes due to small sea level fluctuations. Although the interpretation is based on limited data, a simple depositional model proposed is to consider the algal head boundstones as typical of shallow subtidal to lower intertidal subenvironments. The carbonate succession bearing laminated mat deposits are correlated with low energy intertidal lithofacies, whereas the heterolithic intervals with minor microbial mat intercalations, is interpreted as a high energy intertidal lithofacies. The location of them in the succession is also coherent with the fact that microbially induced sedimentary structures may correlate with turning points of regression-transgression (Noffke 2009) or marine flooding surfaces, representing a period of sediment starvation and non-deposition following a transgression (Mata and Bottjer 2009).

In Fig. 3.25, a section correlated along an E–W transect is displayed. The deepest facies are well exposed to the west, and the shallowest facies are thicker towards the east. The spatial facies distribution is presented in Fig. 3.26. Microbial mats and quartz arenites are dominant to the east, while wavy laminated microbial mats and mudstones predominate in middle areas. Towards deeper areas of the platform, algal head boundstones and mudstones are more abundant. These lithofacies show basal boundaries defined by a relatively abrupt shift from ramp to coastal facies, where tidal influence is well represented by sedimentary structures and biologically influenced mineralization like flat-laminated microbial ecosystems associated to quartz grains (Noffke 2009). Studying modern MISS, Noffke (2007)

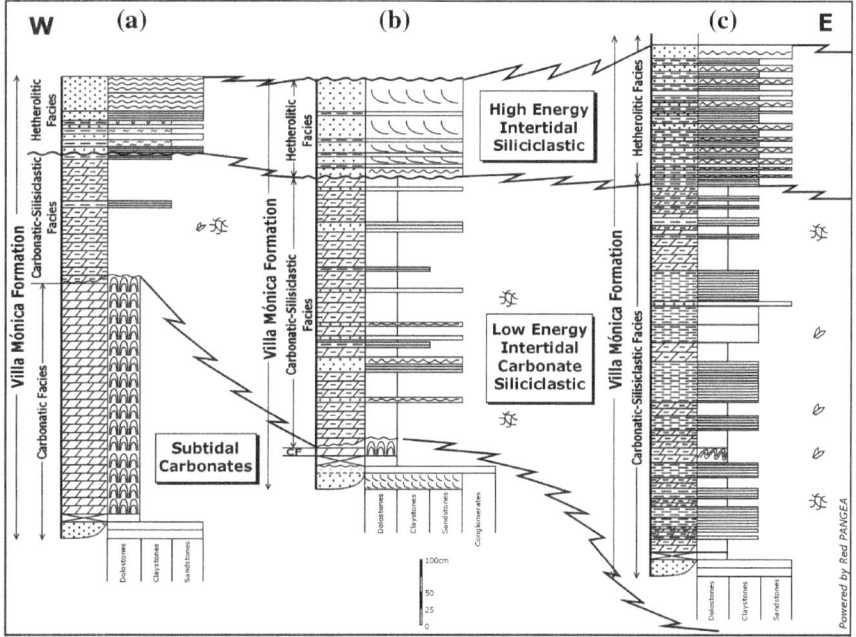

**Fig. 3.25**  Section correlation along an E-W transect. The deepest lithofacies are well exposed to the west, and the shallow lithofacies are thicker towards the east. Taken from Manassero et al. (2012)

## DEPOSITIONAL MODEL FOR THE MIXED FACIES OF THE VILLA MONICA FORMATION

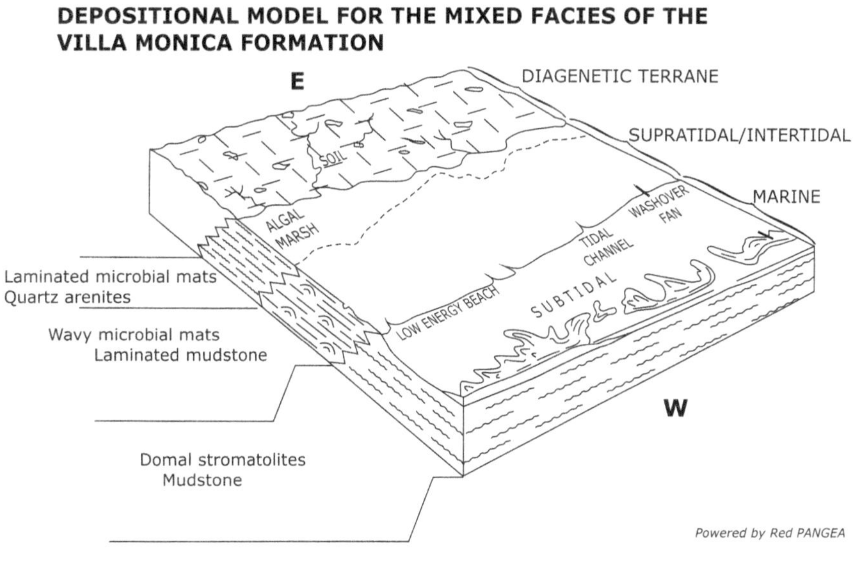

**Fig. 3.26**  Peritidal block diagram model. Quartz arenites and mudstones are dominant to the east (heterolithic facies), while wavy laminated algal mats and mudstones (carbonate-siliciclastic facies) predominate in middle areas. Towards the west and to the deeper areas of the platform, algal head boundstones are developed (carbonate facies). Taken from Manassero et al. (2012)

previously explained how this mineralization takes place: "In modern environments, microbial mats decompose and mineralize the organic matter of the filaments of cyanobacteria and trichomes. The resulting chemical compounds eventually react with ions from the surrounding seawater, and initially, amorphous gels are formed. Later those gels crystallize to form e.g., aragonite, tenorite (the precursor of pyrite), or other minerals such as iron oxides and iron hydroxides. That is those minerals replaced the organic matter of the filaments and trichomes".

In conclusion, at the Sierra La Juanita, the Villa Monica Formation studied units are mainly composed of well-preserved boundstones with columnar stromatolites and of microbial mats associated with siliciclastics, like illitic clays and quartzose sandstones. The mixed-siliciclastic facies, considered by previous authors as exclusively siliciclastic (and named "ferruginous clays") were deposited in intertidal environments. The microunits described as siliciclastic-microbial mat units constitute a unique character, only appreciated on thin sections, of terrigenous translucent quartz grains capable of retaining bacteria filaments in a photosynthetic environment and being stabilized by their cyclic activity. Towards the top of the Villa Mónica Formation, thin sedimentary cycles were formed in a littoral environment, where detrital illitic clays were derived from the erosion of basement rocks. All these subenvironments were developed in a Neoproterozoic shallow epeiric sea with intense blue-green algae production and where prograding peritidal carbonate precipitation took place with minor intercalated siliciclastic events where terrigenous input was dominant. This succession is capped by high-energy coastal siliciclastic sediments represented by quartzose sands.

The concluding interpretation of Manassero et al. (2012) is a simple depositional facies belt with a N–S coastal line where the algal head boundstones and mudstones located to the west are part of the subtidal lithofacies. To the east, the mixed-siliciclastic succession bearing laminated microbial deposits are correlated with low-energy intertidal deposits whereas the heterolithic deposits are disposed also to the most eastern areas, and are interpreted as high-energy intertidal facies with quartzose sandy input. This actual evidence of the presence of MISS was decisive in the evaluation of the paleoenvironmental conditions and also confirmed the rightness of considering the heterolithic facies as part of the Villa Mónica Formation.

### 3.3.6 Age of the Villa Mónica Formation

The Villa Monica Formation age is considered to be Riphean, on the basis of the type of stromatolites (Poiré 1993), and also on the basis of Rb–Sr dating of illitic greenish clays associated with dolostones at Sierras Bayas locality, resulting an age of 793 ± 32 Ma for the burial diagenesis of this unit (Cingolani and Bonhomme 1982). Nevertheless, it must be taken into account that these sediments, apart from illite $1M_1$ (diagenetic) also contain detrital illite ($2M_1$) and ISII derived from illite $2M_1$.

### 3.3.7  Technological Properties of the Clays of the Villa Mónica Formation

Illitic, yellow (or green) clays of the Villa Mónica Formation in the Sierra La Juanita present plasticity index values ranging between 10 and 13, while red clays have plasticity index of 43. These latter high values of plasticity are due to the clay mix between the yellow ones (with plasticity similar to the rest of the clays of the Buenos Aires province deposits) with the red clay of kaolinitic-smectitic composition, mix that can be used as plastic additive in the manufacture of red pottery. Iron oxides not only bring color to the paste, but they also lower its melting point, vitrifying completely at 1200 °C. The mix that takes place in the bulk extraction of the yellow clay with the red one results in a material with optimal characteristics for molding. At the La Placeres quarry, the material did not show the same fittingness due to the presence of silty lenses with high percentages of quartz.

Mechanical properties make these clays suitable for use in the composition of ceramic bodies, since they have very good compressive and flexural strength, both dry and cooked. These clays are used as plasticizer ingredient in paste tiles in the Buenos Aires province, where more than 15000 tons/year are utilized. This material was exploited until recently at the La Siempre Verde and Don Camilo quarries and it is used in the manufacture of tiles known commercially as LA-AM clay (Domínguez and Schalamuk 1999; Alló 2001). From the above mentioned, clays from the Villa Mónica Formation, at Sierra La Juanita, are classified as plastic clays.

## 3.4  Cuchilla de Las Aguilas and Sierra de La Tinta Sector

### 3.4.1  Sedimentary Deposits; Characteristics, Mineralogical, and Chemical Composition

In the area of the Cuchilla de Las Aguilas and Sierra de La Tinta, 7 km to the West of the Sierra La Juanita, and 3 km west of the town of Barker (see Fig. 1.1), the crystalline basement rocks are composed of tonalites, which are exposed at the base of this hill and also at the base of the Cerro El Sombrerito (see Fig. 1.2). These rocks occur as greenish, weathered, and friable material. The sedimentary sequence overlying the basement rocks is represented by the Sierras Bayas Group (here consisting of the Villa Mónica Formation and the overlying Cerro Largo Formation) covered by the Las Aguilas Formation and the latter, in turn, by the Balcarce Formation, both separated by erosional erosive unconformities (Fig. 3.27). Lateral variations of the Villa Mónica Formation are represented here by friable, yellowish quartzites of only 4 m thick overlying the Cerro Largo Formation instead of being composed of clays and/or dolostones, as in other sectors. The Las Aguilas Formation was subject of numerous studies by some of the authors of this piece of work over several years of research (e.g., Zalba 1979, 1982, 1988) Zalba et al.

**Fig. 3.27** Las Aguilas
Formation: Middle
Lithofacies (*1*) and Upper
Lithofacies (*2*), separated by
an erosive unconformity of
the overlying Balcarce
Formation (*upper white line*)

1982, 1988, 2007a, b; Andreis and Zalba 1989; Iñíguez et al. 1989). Figure 3.28 is a stratigraphic section of this formation which consists of three lithofacies, with measured thicknesses ranging from 3–20 m maximum.

The Balcarce Formation, in this sector, is very thin (up to 10 m) and is covered, at the same time, by thin modern deposits (Quaternary). The Balcarce Formation (predominant quartzites with intercalated kaolinitic clay levels) has an important development in the Chapadmalal-Balcarce-Necochea Sector and so, it will be treated extensively in Chap. 6.

The Las Aguilas Formation, outcropping in the Cuchilla de Las Aguilas, was divided from base to top into three lithofacies (Zalba et al. 1988, 2007a): breccias (Lower Lithofacies), claystones and mudstones (Middle Lithofacies) and alternating claystone, siltstone, and sandstone sediments (Upper Lithofacies). The quartzitic sediments overlying the Upper Lithofacies were originally considered as the base of the Balcarce Formation (Zalba et al. 1982). Apparent lenses of alunite $(KAl_3 (SO_4)_2 (OH)_6)$, −0.05 to about 0. 25 m thick and hundreds of meters in lateral extent—occur along sedimentary discontinuities, either at the contact between the Middle Lithofacies (Fig. 3.29a) and the interbedded claystone-siltstone-sandstone deposits of the Upper Lithofacies or, in lesser amounts, intercalated in the first 0.60 m of the Upper Lithofacies (Fig. 3.29b). In the contact between the Middle and the Upper Lithofacies, alunite occurs within, or at the base of bleached clay levels (light gray to white colored) 1–2 m thick. Petrographic observations indicate that the alunite crystals present a columnar habit, typical of crystallization at the walls of open spaces and of a continuous growth, according to the principle of the geometrical selection generally invoked for the occurrence of drusy crystals (Grigor'ev 1965).

The Middle Lithofacies is composed of massive or laminated claystones with minor interbedded sandstones and siltstones. These sediments are red colored at the base, brownish-red at the middle section, and greenish-gray to yellowish-gray at the top. Several lenses of silica nodules (now quartz) were observed in the upper part of the bleached clays containing the alunite deposits of the Upper Lithofacies (Fig. 3.29c). Alunite, part of the kaolinite, as well as diaspore, halloysite, and

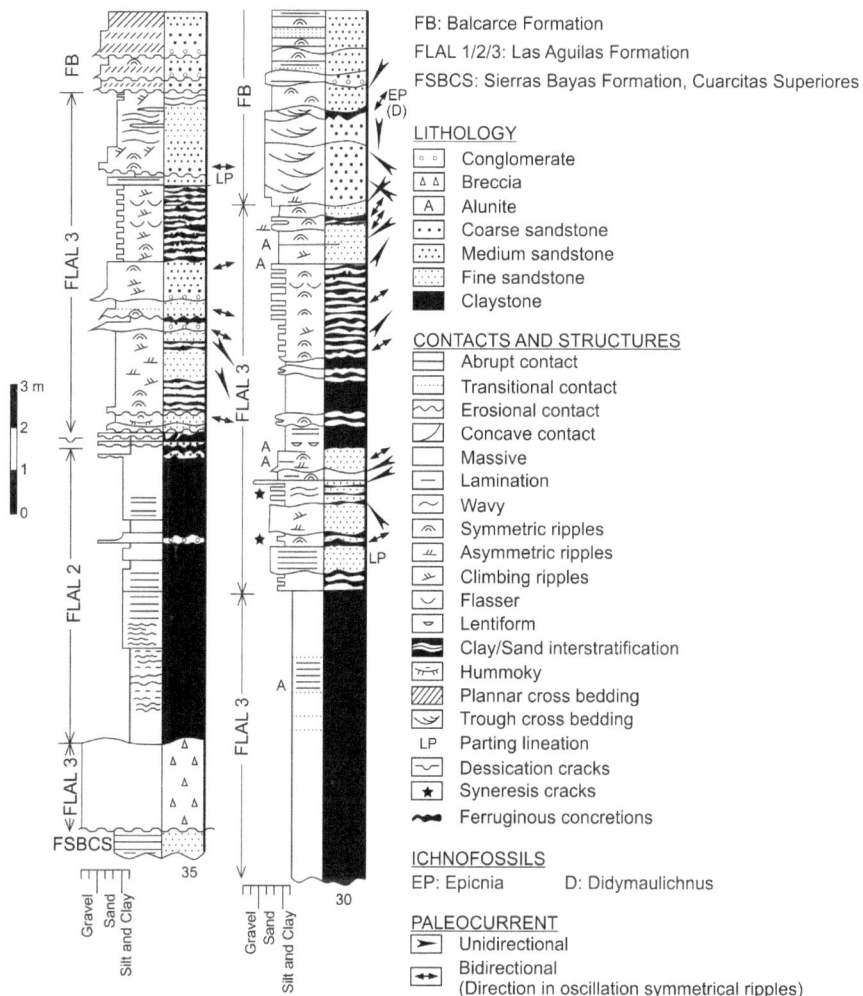

**Fig. 3.28** Stratigraphic section of the Las Aguilas Formation, cover unconformably by quartzites of the Balcarce Formation, Cuchilla de Las Águilas. Taken from Zalba et al. (1988)

goethite, are considered to be diagenetic mineral phases (Zalba 1982, 1988; Iñiguez et al. 1989). The basal reddish-clay sediments contain a high concentration of hematite, and numerous cubic pseudomorphous of hematite after pyrite have been recognized in the greenish-gray to yellowish-gray-clay levels (Fig. 3.29d). Alunite samples analyzed (Zalba et al. 2007a) correspond to the top of the Middle Lithofacies.

X-ray diffraction patterns of the clay material forming the above-mentioned sedimentary layers with different colors indicate significant mineralogical changes. Major amounts of kaolinite and variable proportions of pyrophyllite are present, although the latter decreases upwards. Brownish-red and yellowish-gray clay-rich

**Fig. 3.29** Las Aguilas Formation, Cuchilla de Las Aguilas. **a** Alunite lenses along sedimentary discontinuities. **b** Alunite lenses on the base of the Upper Lithofacies. **c** Lenses of silica nodules in the upper part of the bleached clays containing the alunite deposits of the Upper Lithofacies. **d** Abundant cubic pseudomorphous of hematite after pyrite (Pi), in the *greenish-gray* to *yellowish-gray-clay* in the *top* of the Middle Lithofacies. Taken from Zalba et al. (2007a)

sediments contain illite and ISII (with <15 % smectite), which decrease upwards. Alunite is associated with diaspore, halloysite, and goethite. Variable quartz with overgrowths and very scarce feldspars are present.

Unfortunately, the alunite deposits cannot be exploited because of their lack of vertical development. However, their presence could increase, locally, the content of $Al_2O_3$ and $SO_3$. X-ray diffraction analyses of some selected samples of the Middle Lithofacies are presented in Table 3.14. The corresponding chemical analyses are detailed in Table 3.15.

The pyrophyllite, quartz, feldspars, and part of the kaolinite are considered to be of detrital origin and derived from the weathering, transport, and deposition of previously hydrothermalized crystalline basement rocks (gneisses) cropping out in the San Manuel area (60 km E from the Cuchilla de Las Aguilas) as stated by Zalba (1982), Zalba et al. (1988), and Zalba and Andreis (1998).

Detrital pyrophyllite in the Las Aguilas Formation is shown in Fig. 3.30. Pyrophyllite is also found in clasts of silicified basal breccias of the Las Aguilas Formation at Barker area, considered as reworked calcareous platform sediments. The fabric of these deposits presents abundant silt to fine sand-sized allochems (including ooids, oolites of concentric fabric, and intraclasts) but also shows wavy-crinckled lamination attributed to microbial processes and similar to fabrics observed in modern and ancient mounds that Riding (2000) has termed agglutinated stromatolites. The deposits are classified as ooid grainstones. The oolites may show nucleus of quartz, or show no nucleus (Fig. 3.31a, b). Detrital quartz (center of

**Table 3.14** Mineralogical composition by X-ray diffraction of some selected samples of the Middle Lithofacies of the Las Aguilas Formation

| Sample | Kaolinite | Pyrofilite | Illite | Chlorite | Quartz | Feldspars |
|---|---|---|---|---|---|---|
| 6 | 42 | 33 | 25 | – | Scarce | Scarce |
| 12 | 25 | 15 | 50 | – | Scarce | Scarce |
| 25 | 60 | 20 | 20 | – | Scarce | Scarce |
| 35 "B" | 48 | 20 | 25 | 7 | Scarce | Scarce |
| 36 | 45 | 45 | 10 | – | Scarce | Scarce |

Samples with kaolinite may contain halloysite, undifferentiated by XRD. Taken from Zalba (1979)

**Table 3.15** Chemical composition (weight percent) of the clays of the Las Aguilas Formation

| Sample | $SiO_2$ % | $Al_2O_3$ % | $Fe_2O_3$ % | $TiO_2$ % | CaO % | MgO % | MnO % | $Na_2O$ % | $K_2O$ % | $H_2O$ – % | $H_2O$ +% |
|---|---|---|---|---|---|---|---|---|---|---|---|
| 6 | 48.6 | 34.8 | 2.00 | 0.50 | 0.03 | 0.16 | – | 0.64 | 3.40 | 1.05 | 8.10 |
| 12 | 46.0 | 34.5 | 2.00 | 1.50 | 0.20 | 0.60 | – | 0.76 | 7.22 | 1.08 | 6.30 |
| 25 | 45.9 | 39.2 | 0.80 | 1.70 | – | 0.30 | – | 0.23 | 2.76 | 0.75 | 8.96 |
| 35 "B" | 35.7 | 38.1 | 11.2 | 0.42 | 0.05 | – | 0.97 | 0.39 | 1.95 | 1.54 | 11.1 |
| 36 | 18.1 | 12.1 | 58.0 | 0.98 | 0.05 | 0.04 | 0.71 | 0.40 | 0.89 | 1.73 | 8.76 |

Taken from Zalba (1979)

**Fig. 3.30** Scanning electron micrograph of detrital pyrophyllite in clay deposits of the Las Aguilas Formation, Cuchilla de Las Aguilas, Barker. Note characteristic straight borders. Taken from Zalba (1982)

Fig. 3.31b) shows syntaxial overgrowths. Compaction has put oolites in contact but they are slightly deformed (Fig. 3.31a, b). It is clearly observed that pyrophyllite has grown "in situ" at expense of detrital quartz and kaolinite in fine-grained silica cement in the same sample (Fig. 3.31c). Pyrophillyte has also grown as a selective replacement of oolite nuclei (Fig. 3.31d). Furthermore, agglutinated mammelar

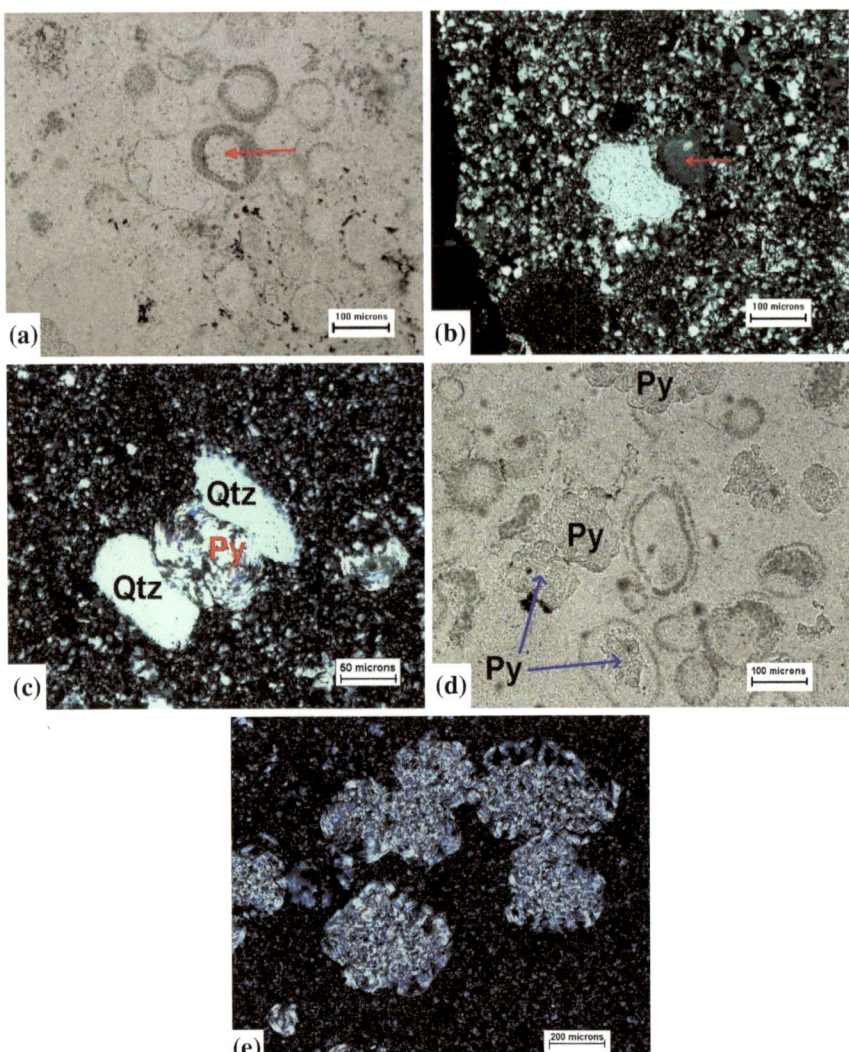

**Fig. 3.31** Photomicrographs of ooid grainstone deposits, Las Aguilas Formation, Cuchilla de Las Aguilas, Barker. **a** Oolites with quartz nucleus (*arrow*). Slight deformation by compaction (PPL). **b** Quartz nucleus (*arrow*). Detrital quartz (*center* of the photograph) with dissolved syntaxial overgrowths in contact with an oolite and with signals of compaction (XPL). **c** Diagenetic pyrophyllite (Py) growing at expense of detrital quartz (Qtz). Note dissolution of syntaxial quartz overgrowths (XPL). **d** Pyrophyllite (Py) growing as diagenetic masses and also as nucleus of oolites (*arrows*) (PPL). e) Mammelar and agglutinated diagenetic pyrophyllite in the same sample (XPL). Taken from Zalba et al. (2010b)

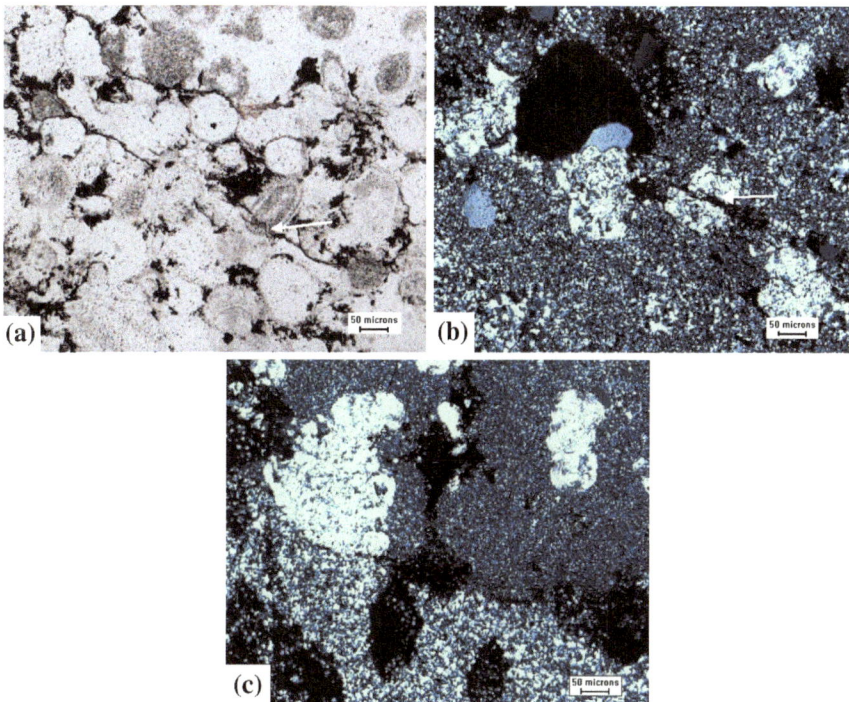

**Fig. 3.32** Photomicrographs of ooid grainstone deposits, Las Aguilas Formation, Cuchilla de Las Aguilas, Barker: **a** Stylolite (*arrow*) affecting grainstone deposits: Ooids have been fractured and dissolved (PPL). **b** Stylolites infilled with hematite cutting diagenetic pyrophyllite (*arrow*) in silicified ooid grainstones (XPL) **c** Two generations of silica cement in contact. Diagenetic pyrophyllite grows only in the micro silica cement (XPL). Taken from Zalba et al. (2010b)

pyrophyllite can be seen in the same deposits (Fig. 3.31e). It is not well understood about the origin of this pyrophyllite but textural relationship with quartz makes a diagenetic origin a reasonable hypothesis for this example. Stylolitization has also affected grainstone deposits of the Las Aguilas Formation, where ooids (Fig. 3.32a) and presumably diagenetically formed pyrophyllite (Fig. 3.32b) have been fractured and dissolved. Chalcedony cement has replaced all previous carbonate cement within the ooid grainstone. It is possible to recognize two different grain sizes of silica cement-textures (Fig. 3.32c) which could be the result of selective recrystallization or replacement of rocks of different porosity. Irregular masses of presumably diagenetic pyrophyllite have grown only in the fine-grained cement.

The Las Aguilas Formation clay deposits have been largely industrially exploited and were considered as lateral facies of the Cerro Negro Formation (the latter found at the Villa Cacique Sector and at the Sierras Bayas Sector) by Iñiguez et al. (1989) and also correlated with the Olavarría Formation by Poiré and Spalletti (2005). The stratigraphic position of the Las Aguilas Formation is still on debate because no definite geologic relationship with the Loma Negra Formation could be sustained until now. We think, from the work of Iñíguez et al. (1989) on, that the Las Aguilas Formation could, eventually, be only correlated with the Cerro Negro Formation. Our reasons for this assumption are

- The base of the Las Aguilas Formation is a silicified limestone breccia with intraclasts, ooids, and oolites inherited from former reworked calcareous deposits and represents an unconformity (paleosurface). The only previous limestone deposits known are those represented by the Loma Negra Formation (see its stratigraphic position in Table 1.1).
- The clay deposits of the Olavarría Formation underlie the Loma Negra Formation (as it is proved at Villa Cacique and Sierras Bayas sectors) and bear no carbonatic deposits.
- The roof of the Loma Negra Formation, in Villa Cacique Sector, is a paleosurface with phosphorites (Leanza and Hugo 1987), which signifies the existence of an important hiatus in the depositional history (erosion) of the basin; so it is not unrealistic to assume that the Loma Negra Formation and the Olavarría Formation could have also been eroded at Cuchilla de Las Aguilas.

### 3.4.2  Scanning Electron Microscopy and Microprobe Analysis of the Middle Lithofacies of the Las Aguilas Formation

SEM performed on clays of the Middle Lithofacies shows textures of face to-face, edge-to-face disposition and "swirl" pattern, typical of sedimentary clay deposits (Fig. 3.33a). Figure 3.33b is a "wet grass" texture, where the clay is, fundamentally, halloysite in tubes. In Fig. 3.33c the association kaolinite (hexagonal crystals) halloysite (tubular crystals) and alunite (pseudocubic crystals) is present. Levels with the highest concentration of alunite are exemplified in Fig. 3.33d, while, in Fig. 3.33e it is possible to observe a geometric arrangement of diaspore crystals in predominantly kaolinitic clays, with associated halloysite and alunite. Detrital pyrophyllite (see Fig. 3.30) shows characteristic straight borders at right angles.

According to Zalba et al. (2007a) SEM and microprobe analyses performed on uncovered polished sections containing massive alunite allowed the authors to identify disseminated crystals of aluminum phosphate sulfate (APS) in the clay matrix (Fig. 3.34a) whose microprobe analyses in percentages ($Al_2O_3$: $30.57 \pm 0.59$; $P_2O_5$: $22.58 \pm 0.2$; $SO_3$: $5.89 \pm 0.17$; $SrO$: $8.33 \pm 0.62$; $CaO$: $1.23 \pm 0.06$: $Ce_2O_3$: $14.86 \pm 0.51$) fall in the compositional field of a solid solution between svanbergite ($SrAl_3(PO_4, SO_4)(OH)_6$) and Ce-florencite ($CeAl_3(PO_4)_2(OH)_6$), two APS minerals of the beudantite and the crandallite groups, respectively (Gaboreau et al. 2005).

SEM images obtained in backscattering mode indicate that alunite has preferentially crystallized in the porous parts of the clay sediments (Fig. 3.34b).

**Fig. 3.33** Scanning electron micrographs of the Middle Lithofacies of the Las Aguilas Formation: **a** Textures of face-to-face, edge-to-face disposition and "swirl" pattern. **b** Tubular halloysite (Hy) "in wet grass" texture. **c** Association of hexagonal kaolinite (K), tubular halloysite (Hy) and pseudocubic alunite (A). **d** Levels with the highest concentration of alunite. **e** Geometric arrangement of diaspore (D) crystals in predominantly kaolinitic clays. Taken from Zalba (1982)

(a)

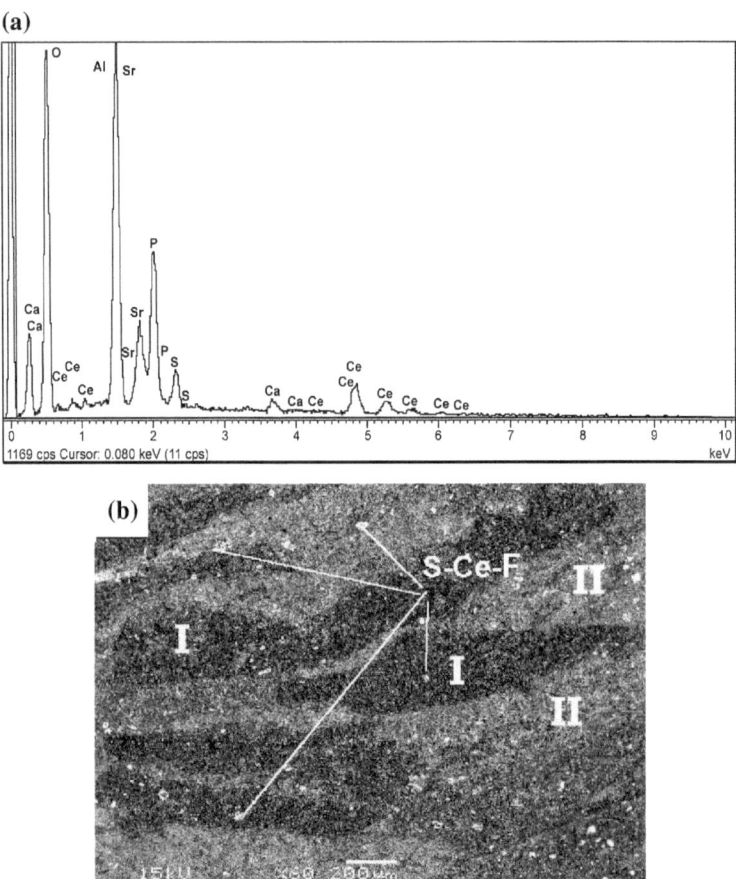

(b)

Fig. 3.34  a  Microprobe  analysis  of  APS  minerals:  S-Ce-F  (svanvergite-Ce-florencite).
b Backscattered electron image of clay deposits bearing APS minerals. (I) High-porosity zones
with alunite, halloysite, diaspore, and svanbergite–Ce–florencite. (II) Low-porosity zones rich in
pyrophyllite, kaolinite, micas, svanbergite–Ce–florencite, and heavy metals. Middle Lithofacies,
Las Aguilas Formation, Cuchilla de Las Aguilas, Barker area. Taken from Zalba et al. (2007a)

## 3.4.3   K-Ar Isotopic Analyses of Alunite

K–Ar dating was carried out on three selected alunite samples (Table 3.16). The
purity of the samples was checked by XRD, which indicates that alunite is mainly
mixed with halloysite and/or kaolinite, two clay minerals which do not contain
potassium. The K-Ar results were plotted on a 40 Ar/36 Ar versus 40 K/36 Ar
isochron diagram (Fig. 3.35). The advantage of using the isochron technique is that
the K-Ar age calculation for a set of data points fitting a line and the determination of
the corresponding initial 40 Ar/36 Ar of the representative samples is made without

**Table 3.16** K–Ar data for three alunite samples containing impurities of kaolinite and halloysite

| Samples | $K_2O$ (%) | $^{40}Ar$ rad ($10^{-6}$ $cm^3$/g STP) | $^{40}Ar$ rad (%) | $^{40}Ar/^{36}Ar$ | $^{40}K/^{36}Ar$ | Age in Ma ($\pm 2\sigma$) |
|---|---|---|---|---|---|---|
| Alu 1 | 0.78 | 7.06 | 23.50 | 386 | 5561 | 261 (13) |
| Alu 2 | 0.77 | 7.12 | 22.90 | 383 | 5259 | 266 (15) |
| Alu 3 | 0.78 | 7.06 | 26.40 | 402 | 6489 | 261 (18) |

Middle Lithofacies, Las Aguilas Formation, Cuchilla de Las Aguilas, Barker area. Alu. 1, 2, 3, samples analyzed (Zalba et al. 2007a)

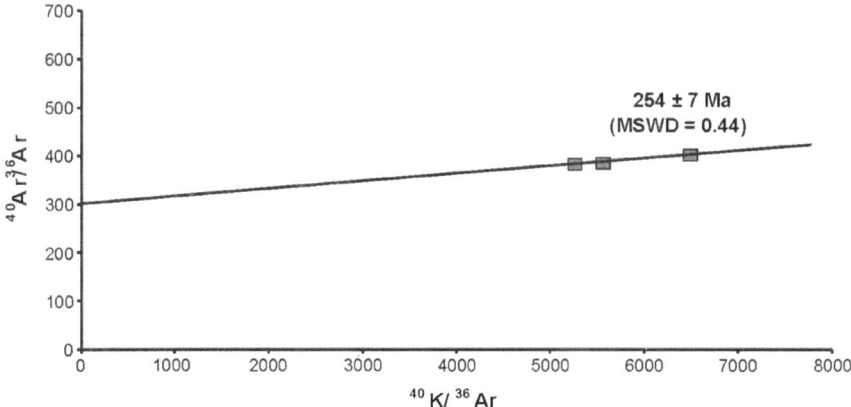

**Fig. 3.35**   40 Ar/36 Ar versus 40 K/36 Ar isochron diagram indicating an age of 254 ± 7 Ma for alunite of the Las Aguilas Formation, Cuchilla de Las Aguilas. Taken from Zalba et al. (2007a)

any assumption. The errors on the slope and intercept depend only on the fitting of the data. In the present case, the regressions were obtained by a Yorkfit regression using ISOPLOT (Ludwig 1991). The K-Ar data for the three samples define a line on an isochron diagram (Fig. 3.34b) with a slope providing a K–Ar date of 254 ± 7 Ma with an intercept at 298.9, which is identical within analytical uncertainty to the measured atmospheric 40 Ar/36 Ar ratio of 300.7. The mean squared weighted deviation (MSWD) is satisfying at 0.44. The mean K-Ar apparent ages of the same size fractions, based on the individual K-Ar values, is 262 ± 19 Ma, which is the same as the intercept age. Taken from Zalba et al. (2007a).

### 3.4.4   *Origin of Alunite and Aluminum Phosphate Sulfate (APS Minerals)*

Dill (2001) reviewed the geology of alunite and aluminum phosphate sulfate (APS minerals) of the Alunite Group and concluded that these minerals occur at low-pH conditions, in a wide range of environments near the surface of the Earth, including weathering, sedimentary, diagenetic, hydrothermal, and also metamorphic and igneous realms. He also stated that peraluminous parent rocks enriched in sulfur and/or phosphorus are a prerequisite for the formation of APS minerals that are stable over a large range of temperature (up to 400 °C at moderately high fluid pressure of up to 1 kbar). Alunite and APS minerals with complex solid-solution series can form according to the local variation of pH and redox conditions in all these environments. Although all of these minerals are stable in acidic conditions, alunite is known to be stable at lower pH than those of APS (Stoffregen and Alpers 1987). Dill (2001) cited many tens of occurrences of alunite and APS minerals in sedimentary environments and noted that these minerals are commonly associated with clay minerals of the Kaolinite Group (kaolinite, halloysite and dickite) and with silica minerals (quartz, chalcedony, opal). The geological process invoked by Dill (2001) for the origin of alunite and APS minerals in aluminous sedimentary rocks is the alteration of an aluminum silicate mineral at low-pH conditions. Such processes of water–rock interaction are initiated by a drastic lowering of the pH of the infiltrating fluids by the release of the sulfuric acid produced by the oxidation of iron sulfides (pyrite, marcasite) in the host rocks. The fact that very abundant pseudomorphous hematite after pyrite crystals has been observed in whitish clays associated with layers where both the alunite and the APS minerals occurred supports this hypothesis. According to the variations of the abundance of mineral phases observed in the different zones, it seems that most of the mineralogical reactions which occurred in the Cuchilla de Las Aguilas area can be summarized by the simplified reaction as follows:

$$2\,\text{illite} + 1\,\text{pyrite} + 1\,H_2O + 3.75\,O_2 \leftrightarrow 1\,\text{alunite} + 1\,\text{kaolinite(or halloysite)} + 5\,SiO_2 + 0.5\,\text{hematite}$$

A similar reaction has already been proposed to interpret the occurrence of alunite in hydrothermally altered limestone and shale sequences hosting the Carlin type gold deposits (Arehart 1996). It should be noted that diaspore can be also a common by-product in this type of hydrothermal reaction which involved meteoric waters as the oxidizing agent (Dill 2001). Aluminum phosphate sulfate minerals commonly crystallize during advanced argillic alteration (Stoffregen and Alpers 1987). However, according to the crystal-chemistry of svanbergite-Ce-florencite solid solution, the APS minerals from the Las Aguilas Formation probably crystallized in less acidic conditions than did alunite in more porous zones. Indeed, the composition of APS is highly dependent on pH. The general trend is decreasing content in S and Sr and increasing P and LREE of APS with increasing pH (Gaboreau et al. 2005). According to this scheme, it seems reasonable to suggest that the APS observed in the less porous zones of the bleached zone represent the alteration products of detrital monazite, coming from the basement rocks and some of the Al-silicate of the sediments (possibly feldspars) by acidic solutions in relatively closed local systems.

Indeed, a low water/rock ratio (due to low porosity and poor renewal of solutions) promoted the high chemical activity of $PO_3^{-3}$ and the rapid increase in pH

**Table 3.17** Paragenetic sequence proposed for the Cuchilla de Las Aguilas, Barker area (Las Aguilas Formation)

| Observations | | Processes | Geological history stage | |
|---|---|---|---|---|
| Minerals | Structures | | | |
| I/S (<15 % Sm) Quartz overgrowth pyrite | Stratification | Compaction organic maturation | Burial stage | |
| | Folding | SW-NE compression | Tectonic effect of Ventania | |
| Kaolinite—halloysite alunite, APS, diaspore hematite, goethite | Pseudomorphous hematite after pyrite Alunite lenses | Invasion of oxidative fluids (meteoric) | Uplift erosion | Telogenesis |

Taken from Zalba et al. (2007a)

conditions necessary to stabilize the APS of the svanbergite-florencite solid solution as the alteration of monazite progressed.

In the Barker area, the identification of oxidized pyrite (pseudomorphous hematite after pyrite) in the Las Aguilas Formation, together with the association of APS minerals (alunite, svanbergite-Ce-florencite) kaolinite, halloysite, diaspore, and goethite, allowed Zalba et al. (2007a) to propose a paragenetic sequence (Table 3.17) based, in part, on Worden and Burley (2003) and Burley and MacQuaker (1992) models. Infiltration of oxidizing fluids was able to strongly dissolve and oxidize the euhedral diagenetic pyrite within the claystones of the Las Aguilas Formation. Basin inversion, uplift, erosion, and infiltration of surficial oxidizing waters in previously more deeply buried sediments belong to the history of most of the sedimentary basins in the world. These phenomena are particularly well developed near the basin margins and are reported in the literature as telogenesis (Worden and Burley 2003). K–Ar dating for the three samples of alunite provides an age of $254 \pm 7$ Ma for the occurrence of the telogenetic stage in the Tandilia basin. According to Burley and Macquaker (1992), the period of time during which the basin-margin sequences can be subjected to telogenesis and hence the intensity of the related alteration processes depends on the timing of structural inversion. We have no reliable data on the origin of the infiltrated oxidizing waters which initiated the advanced argillic alteration in the Las Aguilas Formation. However, all the elements of the above discussion lead us to consider the hypothesis of infiltration of meteoric water as a realistic one. Taken from Zalba et al. (2007a).

## 3.4.5 Microscale Diagnostic Diagenetic Features in the Cuchilla de Las Aguilas

More examples of Microbially Induced Sedimentary Structures (MISS) have been found in the Las Aguilas Formation (Zalba et al. 2010b), described for the first time

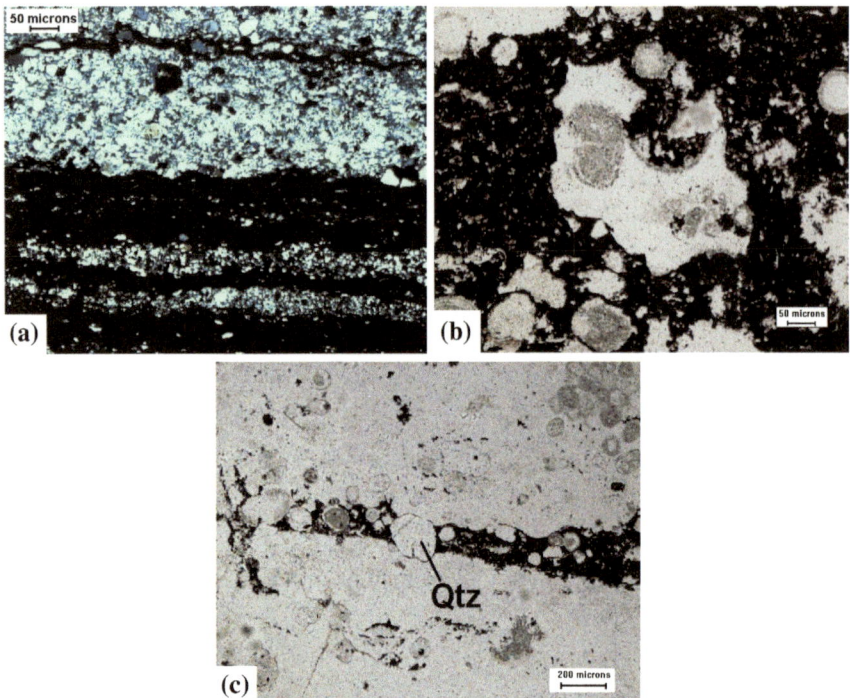

**Fig. 3.36** **a** Microbial mat levels intercalated in ooid grainstone deposits, with quartz grains and ooids trapped within (PPL). **b** Tidal deposits (alternation of clayey and sandy sediments) with microbial mats (*dark*, hematized microbial activity relics) alternating with clastic sediments (XPL). **c** Relics of microbial mats (*dark*) completely hematized with an intraclast (*center* of the photograph) containing ooids. The mats also contain reworked material (clasts and ooids) (PPL). Taken from Zalba et al. (2010b)

as tidal deposits by Zalba (1979) at the Barker area. These structures (Fig. 3.36a) are very similar (hematized, dark microbial mats, wrinkle structures, detrital grains with their long axis parallel to bedding planes trapped within the mats, dome structures) to those described previously for the Villa Mónica Formation. That is why they are considered now as mixed facies and not purely siliciclastic, as they were firstly defined (Zalba 1979). The dark, hematized cement shown in Fig. 3.36b, c, has been attributed to microbial mat relics in silicified ooid grainstones, with whole ooids, fractured ooid intraclasts (Fig. 3.36b, c), and detrital grains trapped within, interpreted as reworked sediments (Fig. 3.36c). The textural relationship between the trapped quartz grains and the mat relics (Fig. 3.36c) is the proof that the hematized deposit is a microbially induced structure and not a fracture infilled with hematite.

During the Permian, uplift of the Ventania System, located 150 km to the SW of Barker, took place (Varela et al. 1985; Von Gosen and Buggisch 1989). According to Zalba et al. (2007a) simultaneous compression from the SW occurred in Tandilia.

Due to basin inversion, a new stage in the geological history of Tandilia began. During this telogenetic stage kaolinite formed due to the introduction of oxidized meteoric fluids related to pyrite oxidation, in the Neoproterozoic Las Aguilas Formation (younger than the Villa Mónica Formation and outcropping 7 km to the east of the Estancia La Siempre Verde), where it is also associated with halloysite, together with diaspore and APS minerals. Based on this evidence it is not unrealistic to suppose that regional telogenetic processes developed in the Barker area also affected the Villa Mónica Formation.

### 3.4.6 Technological Properties of the Clays of the Las Aguilas Formation

Middle Lithofacies: according to numerous technological analyses of our own research projects and depending on their physical and chemical characteristics and coloring ($Fe_2O_3$ content) these clays may be classified as:

1. Suitable for high quality refractory: refractory clays with 43–55 % $SiO_2$; 33–38 % $Al_2O_3$; 0.6–1.2 % $Fe_2O_3$ and a Pyrometric Cone Equivalent (PCE) between 31 and 32.5. The minerals present are kaolinite or halloysite with lower proportions of pyrophyllite, illite and I/S. This kind of clays is very scarce.
2. Suitable for white ceramic, ware, tiles and refractory of medium quality: white plastic clays and green clays with 45–46 % $SiO_2$, 34–36 % $Al_2O_3$, 0.9–2 % $Fe_2O_3$ and with a Pyrometric Cone Equivalent (PCE) between 29 and 26. The clay minerals are illite, kaolinite, pyrophyllite and I/S. These clays are very scarce.
3. Suitable for red ceramics: clays with 8–18 weight % of $Fe_2O_3$; 47 % $SiO_2$, 31–33 % $Al_2O_3$ and a Pyrometric Cone Equivalent (PCE) between 21 and 29. The phylosilicates are kaolinite, pyrophyllite, illite and I/S. Their use in the refractory industry is limited by the percentage of iron. This is the most abundant type of clay in the area.
4. Limited technological use: variable $Fe_2O_3$ content, they contain 55–60 % $SiO_2$, 23–28 % $Al_2O_3$, and with a Pyrometric Cone Equivalent (PCE) between 21 and 23. Pyrophyllite, kaolinite, illite and I/S are present.

In general, clays of the Middle Lithofacies have tenors of $Fe_2O_3$ exceeding those recommended for the preparation of high quality refractory materials (it should not exceed 1 %), so, while $Al_2O_3$ % values are high, they cannot be destined for that purpose. All of them are plastic clays, with refractory and semiplastic varieties. They were intensively exploited and utilized in red ceramic and cement industries and constitute one of the most important resources in terms of reserves and quality.

Upper Lithofacies: the clay sediments of this lithofacies, intercalated with quartzite levels, form layers of reduced individual thickness and, rarely, the whole sequence reaches up to 9 m of total thickness. They are semiplastic, whitish to yellowish claystones and siltstones and are moderately resistant to fire. In general, they are mixed with white pyrophyllitic clays, of high quality, generating a good combination for "porcellanatos" (Domínguez and Ullman 2005).

# References

Ahmad AHM, Bhat GM (2006) Petrofacies, provenance and diagenesis of the dhosa sandstone member (Chari Formation) at Ler, Kachchh sub-basin, Western India. J Asian Earth Sci 27:857–872

Alló W (2001) Los yacimientos de arcillas illíticas ferruginosas La Siempre Verde y La Placeres de Bárker. Tesis Doctoral (inédito), Universidad Nacional del Sur, Bahía Blanca, p 235

Alló W, Domínguez E, Cravero F (1986) Caracterización de la illita del yacimiento La Siempre Verde, Barker, Buenos Aires. Politipos indicadores de un rango termal entre diagénesis profunda y metamorfismo leve. In: Proc 33° Reunión de Mineralogía y Metalogenia, de Brodtkorb MK, Schalamuk, IB (eds.), La Plata, pp 27–35

Andreis RR (2003) The Tandilia system province of Buenos Aires, Argentina: its sedimentary successions. In: Domínguez E, Mas GR, Cravero F (eds) 2001, A clay odyssey, Elsevier, Amsterdam, pp 15–22

Andreis RR, Zalba PE (1989) Estratigrafía y paleogeografía de las secuencias cuarciticas al oeste de Barker (Buenos Aires, Argentina). In: Proc 1° Jornadas Geológicas Bonaerenses, Tandil, pp 909–930

Andreis RR, Zalba PE, Iñiguez Rodriguez AM, Morosi M (1996) Estratigrafía y evolución paleoambiental de la sucesión superior de la Formación Cerro Largo, Sierras Bayas (Buenos Aires, Argentina). In: Proc 6° Reunión Argentina de Sedimentología, pp 293–298

Arehart GB (1996) Characteristics and origin of sediment-hosted disseminated gold deposits: a review. Ore Geol Rev 11:383–403

Awramik SM (1984) ancient stromatolites and microbial mats. In: Cohen Y, Castenholz RW, Halvorson HO (eds) Microbial mats: stromatolites. Liss AR Publisher, New York, pp 1–22

Bouougri E, Porada H (2002) Mat related sedimentary structures in Neoproterozoic peritidal passive margin deposits in the West African Craton (Anti-Atlas). Sed Geol 153:85–106

Brewer R (1960) Cutans: their definition, recognition and interpretation. Eur J Soil Sci 11(2): 280–292

Brewer R (1976) Fabric and mineral analysis of soils. Krieger RE Publishing Co, Huntington, New York, p 480

Burley SD, MacQuaker JHS (1992) Authigenic clays, diagenetic sequences and conceptual diagenetic models in contrasting basin-margin and basin-center North Sea Jurassic sandstones and mudstones. In: Houseknecht DW, Pittman ED (eds) Origin, diagenesis and petrophysics of clay minerals in sandstones, SEPM Special Publication, Tulsa, Oklahoma, USA, vol 47. pp 81–110

Caudill MR, Driese SG, Mora CI (1996) Preservation of a paleo-vertisol and an estimate of Late Mississippian paleoprecipitation. J Sedim Res 66:58–70

Choquette PW, Pray LC (1970) Geologic nomenclature and classification of porosity in sedimentary carbonates. AAPG Bull 54:207–250

Cingolani CA (2011) The Tandilia system of Argentina as a southern extension of the Río de la Plata craton: an overview. Int J Earth Sci 100:221–242

Cingolani CA, Bonhomme MG (1982) Geochronology of La Tinta upper Proterozoic sedimentary rocks Argentina. Precambrian Res 18(1–2):119–132

Dai Y, Song H, Shen J (2004) Fossil bacteria in Xuanlong iron ore deposits of Hebei province. Sci China Ser D Earth Sci 47:347–356

Dill HG (2001) The geology of aluminium phosphates and sulphates of the alunite group minerals: a review. Earth Sci Rev 53:35–93

Domínguez E, Schalamuk I (1999) Recursos minerales de las Sierras Septentrionales, Buenos Aires. In: Zappettini E, (ed) Recursos Minerales de la República Argentina, SEGEMAR, Buenos Aires. Anales, vol 35. pp 183–190

Domínguez E, Ullman R (2005) Arcillas e industria cerámica. In: De Barrio R, Etcheverry R, Caballé, M, Llambías E (eds.) Proc 16º Congreso Geológico Argentino, La Plata, Asociación Geológica Argentina, Buenos Aires, pp 397–408

Driese SG, Foreman JL (1992) Paleopedology and paleoclimatic implications of late ordovician vertic paleosols, southern Appalachians. J Sed Petrol 62:71–83

Driese SG, Mora CI (1993) Physico-chemical environment of pedogenic carbonate formation in devonian vertic paleosols, central Appalachian, USA. Sedimentology 40:199–216

Driese SG, Mora CI, Cotter E, Foreman JL (1992) Paleopedology and stable isotope geochemistry of Late Silurian vertic paleosols, bloomsburg formation, central Pennsylvania. J Sed Petrol 62:825–841

Druschke P, Jiang G, Anderson TB, Hanson AD (2009) Stromatolites in the late ordovician eureka quartzite: implications for microbial growth and preservation in siliciclastic settings. Sedimentology 56:1275–1291

Dunham RL (1962) Classification of carbonate rocks according to depositional texture. Memoir Am Assoc Petrol Geol 1:108–121

Flugel E (2004) Microfacies of carbonate rocks, analysis, interpretation and application. Springer, Berlin, p 976

Gaboreau S, Beaufort D, Viellard Ph, Patrier P, Bruneton P (2005) Aluminium phosphate-sulphate minerals associated with Proterozoic unconformity-type uranium deposits in the East Alligator River Uranium Field, Northern Territories, Australia. The Can Mineral 43:813–827

Garrido L, Zalba PE, Pereira E (1984) Estudio de Yacimientos de Arcilla de El Ferrugo y Constante 10, Provincia de Buenos Aires. II Tecnología. Revista Latinoamericana de Ingeniería, Química y Química Aplicada 14:207–216

Gerdes G, Klenke T, Noffke N (2000) Microbial signatures in peritidal siliciclastic sediments: a catalogue. Sedimentology 47:279–308

Gerdes G, Krumbein WE, Reineck HE (1991) Biolaminations-ecological versus depositional dynamics. In: Einsele G, Ricken W, Seilacher A (eds) Cycles and events in stratigraphy. Springer, Berlin, pp 592–607

Grigor'ev DP (1965) Ontogeny of minerals: Israel program for scientific translations ltd, S. Marson, Jerusalem, p 250

Hemingway IE, Riddler GP (1982) Basin inversion in North Yorkshire. Trans Instn Min Metal 91:175–186

Hofmann HJ (1975) Stratiform precambrian stromatolites, Belcher islands, Canada: Relation between silicified microfossils and microstructure. Am J Sci 275:1121–1132

Iñiguez MA, Zalba PE (1974) Nuevo nivel de arcilitas en la zona de Cerro Negro, partido de Olavarría, provincia de Buenos Aires. LEMIT Serie 2(264):95–100

Iñiguez AM, Del Valle A, Poiré D, Spalletti L, Zalba P (1989) Cuenca Precámbrica-Paleozoico inferior de Tandilia, Provincia de Buenos Aires. In: Chebli G, Spalletti LA (eds) Cuencas sedimentarias argentinas. Instituto Superior de Correlación Geológica, Universidad Nacional de Tucumán, Serie Correlación Geológica, vol 6, pp 245–263

Iñiguez AM, Zalba PE, Andreis RR (1990) Mineralogy and Chemistry of Cambrian (?) paleosols, Tandilia System, Buenos Aires Province, Argentina. In: Farmer VC, Tardy Y (eds) Proc 9º international clay conference 1989, Institut Géologie, Strasbourg, France, Mémoire, vol 85. pp 175–184

Iñiguez AM, Manassero MJ, Poiré DG, Maggi JH (1996) Génesis y procedencia de sedimentitas cuarzosas del area de Olavarría, Provincia de Buenos Aires, Argentina. In: Proc 6º Reunión Argentina de Sedimentología, Bahía Blanca, pp 61–66

Jiang G, Christie-Blick N, Kaufman A, Baner-jees D, Rai V (2003) Carbonate platform growth and cyclicity at a terminal Proterozoic passive margin, Infra Krol Formation and Krol Group, Lesser Himalaya, India. Sedimentology 50:921–952

Kah L, Bartley L, Stagner A (2009) Reinterpreting a Proterozoic enigma: conophyton-Jacutophyton stromatolites of the Mesoproterozoic Atar Group, Mauritania. Spec Publ Int Assoc Sedimentol 41:277–296

Keller WD (1978) Classification of kaolins exemplified by their texture in scan electron micrographs. Clays Clay Miner 26:1–20

Krumbein WE (1994) Biostabilization of sediments. BIS-Verlag, Oldenburg, p 256

Larsen G, Chilingar GV (1979) Diagenesis in sediments and sedimentary rocks. Dev Sedimentol 25:579

Leanza HA, Hugo C (1987) Descubrimiento de fosforitas sedimentarias en el Proterozoico superior de Tandilia, Buenos Aires Argentina. Rev Asoc Geol Argentina 42(3–4):417–428

López K (2006) Estudio geológico, geoquímico, mineralógico y tecnológico de las Arcillas del área de Estancias Araquistain-Viuda de Manson-La Rosalía en Sierras Septentrionales. Tesis Doctoral 902 (inédito). Facultad de Ciencias Naturales y Museo, Universidad Nacional de La Plata, p 245

López K, Botto IL, Etcheverry R (2002) Geología y mineralogía de las arcilitas localizadas en las Estancias La Rosalía, San Eduardo y Sierra de los Barrientos, Provincia de Buenos Aires. In: Brodtkorb de MK, Koukharsky M, Leal P (eds.) Proc 6° Congreso de Mineralogía y Metalogenia, Buenos Aires, pp 239–246

Ludwig KR (1991) ISOPLOT: A plotting and regression program for radiogenic isotope data, version 2.71: U.S. Geological Survey, Open-File Report, 91–445

Manassero MJ, Zalba PE, Morosi M (1986) Neoproterozoic peritidal facies of the Villa Monica Formation, Sierra La Juanita, Tandilia. Rev Asoc Geol Argentina 69(1):28–42

Manassero M (1986) Estratigrafía y estructura en el sector oriental de la localidad de Barker, Provincia de Buenos Aires. Rev Asoc Geol Argentina 41(3–4):375–384

Manassero MJ, Zalba PE, Morosi M (2012) Neoproterozoic peritidal facies of the Villa Monica Formation, Sierra la Juanita, Tandilia. Rev Asoc Geol Argentina 69(1):28–42

Mata S, Bottjer D (2009) The paleoenvironmental distribution of Phanerozoic wrinkle structures. Earth Sci Rev Microb Mats Earth's Fossil Rec Life Geobiol 96(3):181–195

Morton N (1987) Jurassic subsidence history in the Hebrides, NW Scotland. Marine Petroleum Geol 4:226–242

Noffke N (2006) Spatial and temporal distribution of microbially induced sedimentary structures: a case study from siliciclastic storm deposits of 2.9 Ga old Witwatersrand Supergroup, South Africa. Precambrian Res 146:35–44

Noffke N (2007) Microbially induced sedimentary structures in Archean sandstones: a new window into early life. Gondwana Res 11:336–342

Noffke N (2009) The criteria for biogeneicity of microbially induced sedimentary structures (MISS) in Archean, sandy deposits. Earth Sci Rev Microb Mats Earth Fossil Rec Life Geobiol 96:173–180

Noffke N, Gerdes G, Klenke T, Krumbein WE (1997) A micrsocopy sedimentary succession of graded sand and microbial mats in modern siliciclastic tidal flats. Sedimentary Geol 10:1–6

Noffke N, Gerdes G, Klenke T, Krumbein WE (2001) Microbially induced sedimentary structures —a new category within the classification of primary structures. J Sedimentary Res 71:649–656

Noffke N, Hazen R, Nhleko N (2003) Earth's earliest microbial mats in a siliciclastic marine environment (Mozaan Group, 2.9 Ga, South Africa). Geology 31:673–676

Noffke N, Eriksson KA, Hazen RM, Simpson EL (2006) A new window into early archean life: microbial mats in Earth's oldest siliciclastic tidal deposits (3.2 Ga Moodies Group, South Africa). Geology 34:253–256

Pevear DR (1999) Illite and hydrocarbon exploration. Proc Natl Acad Sci U.S.A. PMCID: PMC34286 Colloquium Paper 96(7):3440–3446

Poiré DG (1987) Mineralogía y sedimentología de la Formación Sierras Bayas en el núcleo septentrional de las sierras homónimas, Olavarría provincia de Buenos Aires. Tesis Doctoral 494 (inédito). Facultad de Ciencias Naturales y Museo, Universidad Nacional de La Plata, p 271

Poiré DG (1993) Estratigrafía del Precámbrico sedimentario de Olavarría Sierras Bayas, provincia de Buenos Aires, Argentina. In: Proc 13° Congreso Geológico Argentino y 3° Congreso de Exploración de Hidrocarburos, Mendoza, vol 2. pp 1–11

Poiré DG, Iñíguez AM (1984) Miembro Psamopelitas de la Formación Sierras. Bayas, Partido de Olavarría, Provincia de Buenos Aires. Rev Asoc Geol Argentina 39:276–283

Poiré DG, Spalletti LA (2005) La cubierta sedimentaria Precámbrica-Paleozoica inferior del Sistema de Tandilia. In: de Barrio RE, Etcheverry RO, Caballé MF, Llambías E (eds) Geología y Recursos Minerales de la Provincia de Buenos Aires. Proc 16° Congreso Geológico Argentino, Relatorio 4, La Plata, pp 51–68

Pratt BR, James NP (1986) The St George Group (Lower Ordovician) of western Newfoundland: tidal flat island model for carbonate sedimentation in shallow epeiric seas. Sedimentology 33:313–343

Reineck HE, Singh IB (1986) Sedimentary depositional environments. Springer, Berlin, p 549

Riding G (2000) Microbial carbonates: the geological record of calcified bacterial-algal mats and biofilms. Sedimentology 47:179–214

Riding R (2011) The nature of stromatolites: 3500 million years of history and a century of research. Lectures in earth sciences, Springer, vol 131. pp 29–74

Schauer C, Venier J (1967) Observaciones geológicas en la zona de Barker, Sierra de la Tinta, Provincia de Buenos Aires. Notas de la Comisión de Investigaciones Científicas, Provincia de Buenos Aires 5(6):1–18

Schieber J (1998) Possible indicators of microbial mat deposits in shales and sandstone: examples from the mid-proterozoic belt supergroup, Montana, USA. Sedimentary Geol 120:105–124

Schieber J (2007) Oxidation of detrital pyrite as a cause for marcasite formation in marine lag deposits from the Devonian in the eastern US. Deep-Sea Res 54:1312–1326

Schieber J, Riciputi L (2004) Pyrite and marcasite coated grains in the ordovician winnipeg formation, Canada: an intertwined record of surface conditions, stratigraphic condensation, geochemical reworking and microbial activity. J Sed Res 75:907–920

Scotchman IC (1991a) The geochemistry of concretions from the kimmeridge clay formation of southern and eastern England. Sedimentology 38:79–106

Scotchman IC (1991b) Kerogen facies and maturity of the kimmeridge clay formation of Southern and Eastern England. Marine Petroleum Geol 8:278–295

Stoffregen R, Alpers Ch (1987) Svanbergite and woodhouseite in hydrothermal ore diposits: implications for apatite destruction during advanced argillic alteration. Canad Mineral 25:201–212

Varela R, Dalla Salda L, Cingolani C (1985) La edad Rb-Sr del Granito de Vela, Tandil. In: Proc 1° Jornadas Geológicas Bonaerenses (Tandil) Comisión Investigaciones Científicas, provincia de Buenos Aires, La Plata, Argentina, pp 881–891

Von Gosen W, Buggisch W (1989) Tectonic evolution of the Sierras Australes fold and thrust belt (Buenos Aires province/Argentina). Geologisches Rundschau 79(3):797–821

Walter MR (1994) Stromatolites: the main geological source of information on the evolution of the early Benthos. In: Bebgstone S (ed) Early life on earth nobel symposium. Columbia University Press, New York, pp 270–286

Walter MR, Bauld J, Des Marais DJ, Schopf JW (1992) A general comparison of microbial mats and microbial stromatolites: bridging the gap between the modern and the fossil. In: Schopf JW, Klein C (eds) The Proterozoic biosphere: an interdisciplinary study. Cambridge University Press, New York, pp 335–338

Williamson WO (1980) Experiments relevant to the genesis of clay mineral orientation in natural sediments. Clay Miner 15:95–97

Worden RH, Burley SD (2003) Sandstone diagenesis: the evolution of sand to stone. In: Burley SD, Worden RH (eds.) Sandstone diagenesis: recent and ancient, Reprint series of the international association of sedimentologists, Blackwell Publishing Ltd., New York, pp 3–44

Zalba PE (1979) Clay deposits of Las Aguilas Formation, Barker, Buenos Aires Province, Argentina. Clay Miner 27(6):433–439

Zalba PE (1981) Nuevo nivel de arcilitas sobre la caliza en la Cantera Loma Negra, Barker. Rev Asoc Geol Argentina 36(1):99–102

Zalba, PE (1982) Scan electron micrographs of clay deposits of Buenos Aires Province, Argentina. International clay conference, Bologna-Pavia, Italy, 1981. Developments in Sedimentology, ol. 35. Elsevier, Ámsterdam, pp 513–528

Zalba, PE (1988) Arcillas de las Sierras Septentrionales de Buenos Aires. Publicación Especial N° 1, CETMIC-CIC, Provincia Buenos Aires, La Plata, p 62

Zalba PE, Andreis RR (1998) Basamento cristalino saprolitizado y secuencia sedimentaria suprayacente en San Manuel, Lobería, Sierras Septentrionales de Buenos Aires, Argentina. In: Proc 7° Reunión Argentina de Sedimentología, Salta, pp 143–153

Zalba PE, Andreis RR (2001) Stratigraphy, sedimentology and mineralogy of Neoproterozoic clay deposits, Sierras de Tandilia, Province of Buenos Aires, Argentina. Economical importance, In: 12th international clay conference, Pre-simposium field trip, Bahía Blanca, p 80

Zalba PE, Andreis RR, Lorenzo F (1982) Consideraciones estratigráficas y paleoambientales de la secuencia basal eopaleozoica en la Cuchilla de Las Aguilas, Barker, Argentina. In: Proc 5° Congreso Latinoamericano de Geología Argentina, vol 2. Buenos Aires, pp 389–409

Zalba PE, Andreis RR, Iñíguez AM (1988) Formación Las Aguilas, Sierras Septentrionales de Buenos Aires, nueva propuesta estratigráfica. Rev Asoc Geol Argentina 43(2):198–209

Zalba PE, Manassero M, Laverret EM, Beaufort D, Meunier A, Morosi M, Segovia L (2007a) Middle Permian telodiagenetic processes in Neoproterozoic sequences, Tandilia System, Argentina. J Sediment Res 77:525–538

Zalba PE, Manassero M, Morosi M (2007b) Meteorización y diagénesis de dolomías estromatolíticas, Formación Villa Mónica (Precámbrico) Sierra de la Juanita, Tandilia. In: Proc 6° Jornadas Geológicas y Geofísicas Bonaerenses. Mar del Plata Abstract, p 46

Zalba PE, Manassero M, Morosi ME, Conconi MS (2010a) Preservation of biogenerated mixed facies: a case study from the Neoproterozoic Villa Mónica Formation, Sierra La Juanita, Tandilia, Argentina. J Appl Sci 10(5):363–379

Zalba PE, Morosi ME, Manassero M, Conconi MS (2010b) Microscale diagnostic diagenetic features in Neoproterozoic and Ordovician units, Tandilia basin, Argentina: a review. J Appl Sci 10(22):2754–2772

# Chapter 4
# Lobería County

**Abstract** San Manuel clay quarries area is located 7 km NW of the town of San Manuel and is a source of important reserves of clay deposits, not exploited at present. Three hills known as Cerro Del Pueblo, Cerro Del Medio and Cerro Reconquista, aligned in a NW–SE direction. Geologically, these hills are composed of fresh and weathered crystalline basement rocks (residual deposits) crowned by thick sedimentary quartzite strata. The rocks of the basement present reddish or purple colorations at the base, with evidence of increasing upward alteration. Going upwards it is possible to distinguish "in situ" relicts of the original rock immersed in argillized zones, followed by completely transformed basement rocks into massive clays: saprolite. Over the weathered basement rocks a siliciclastic sedimentary succession rests unconformably. The sedimentary cover corresponds to the Cuarcitas Inferiores (Villa Mónica Formation). Studies performed by Zalba and Andreis (1998) showed that the Cuarcitas Inferiores wedge east until they almost disappear at Los Cinco Nietos and Julián Luis quarries. This phenomenon hampered for many years its geological correlation with other similar lithofacies of Tandilia. The sequence continues upward with 8 m of claystones, siltstones, and minor quartzites separated by an erosional unconformity from the overlying Cerro Largo Formation. The composition of the phyllosilicates was determined by petrography and X-ray diffraction. SEM shows a typical hydrothermal, locked texture of a mineral formed "in situ". Clay sediments, which cover the residual deposits, lie between quartzite strata (Villa Mónica Formation). The presence of intraclasts of pyrophyllite proves that in the sediments it is detrital and not hydrothermal. MISS have been identified and their preservation prove the absence of hydrothermal processes in these sediments. Technological tests indicate that residual materials are of regular plasticity and are suitable for the manufacture of red and structural ceramics. The presence of pyrophyllite in residual clays increases their refractoriness. Sedimentary clays are suitable for the manufacture of clear ceramics and faience as well as for loading material (rubber). Those with high content of iron oxides could also be used in the preparation of common bricks.

**Keywords** San Manuel · Geology · Stratigraphy · Mineralogy · Residual clays · Sedimentary clays · Pyrophyllite · MISS · Technological properties

© Springer International Publishing Switzerland 2016                                            93
P.E. Zalba et al., *Gondwana Industrial Clays*,
Springer Earth System Sciences, DOI 10.1007/978-3-319-39457-2_4

## 4.1   San Manuel Sector

### 4.1.1   *Residual and Sedimentary Deposits; Characteristics, Mineralogical and Chemical Composition*

San Manuel clay quarries area is located 7 km NW of the town of San Manuel, 50 km E from the town of Barker, and is a source of important reserves of clay deposits, not exploited at present. In San Manuel three hills of low altitude (less than 300 m) are known in the geological literature as Cerro Del Pueblo, Cerro Del Medio and Cerro Reconquista, aligned in a NW–SE direction (Fig. 4.1).

Numerous authors have studied the area: Dristas and Frisicale (1984), Frisicale (1991), Zalba (1988), Zalba and Andreis (1998, 2001), Schalamuk et al. (1992), Fernández et al. (2007), among others. Geologically, these hills are composed of fresh and weathered crystalline basement rocks crowned by thick quartzite strata (Fig. 4.2a, b).

Zalba and Andreis (1998) studies were accomplished in Los Cinco Nietos, Julián Luis, and La Primavera quarries at Cerro Reconquista. According to these authors, the crystalline basement rock (Complejo Buenos Aires) is represented in the area by migmatites (granodiorites) showing weathering evidence. Sub-parallel bands

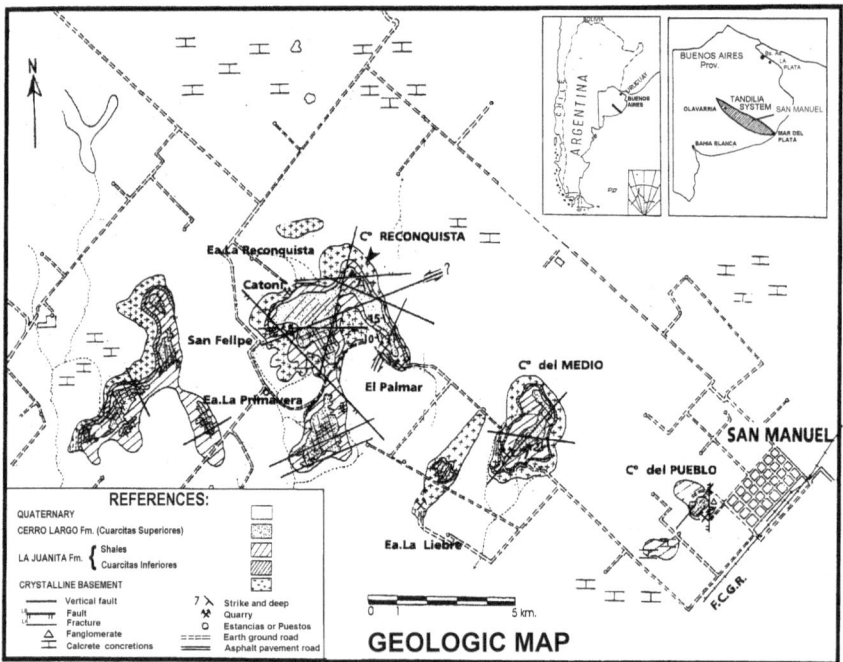

**Fig. 4.1** Geological map of the Cerro Reconquista, Cerro del Medio and Cerro del Pueblo, located west of San Manuel town. Taken from Zalba and Andreis (1998)

**(a)**

**(b)**

Fig. 4.2 **a** San Manuel. Los Cinco Nietos quarry, the mined clay level corresponds to saprolite. **b** Scheme of the Fig. 4.2a, where saprolite, in the weathered basement and the Neoproterozoic sedimentary sequence are observed. La Juanita Formation: Cuarcitas inferiores (A1) and Miembro dolomías (A2), Cerro Largo Formation (B). Taken from Zalba and Andreis (1998)

(foliation) and rare ptigmatic folds, which characterize the less-weathered rock, disappear upwards due to saprolitization (weathering) processes and migmatites acquires a homogeneous and solid appearance.

This process is mainly visible in the Cerro Reconquista, Los Cinco Nietos, and Julián Luis quarries, due to clay mining. The rocks of the basement, with thicknesses between 10 and 22 m, present reddish or purple colorations at the base of the outcrop, with evidence of increasing upward alteration. Going upwards it is possible to distinguish "in situ" relicts of the original rock immersed in argillized zones, followed by completely transformed basement rocks into massive clays.

Over the weathered basement rocks (Fig. 4.3: CB) a siliciclastic sedimentary succession rests unconformably (Fig. 4.3: A1; A2 and B) with a thickness of up to 40 m (Cerro Reconquista), with subhorizontal disposition or with variable dip (up to 15°) towards the West.

The sediments that partially cover the crystalline basement rocks correspond to the Cuarcitas Inferiores (Villa Mónica Formation) as shown in Fig. 4.3: A1. Studies performed by (Zalba and Andreis 1998) showed that the Cuarcitas Inferiores wedge east until they almost disappear at Los Cinco Nietos and Julián Luis quarries. This phenomenon made difficult the understanding of the geology of San Manuel and, therefore, hampered for many years its geological correlation with other areas of Tandilia. The fact is that existing two quartzite levels in the Sierras Bayas Group, known as Cuarcitas Inferiores and Cuarcitas Superiores, respectively, when one of the two levels is almost absent, as is the case of Los Cinco Nietos and Julián Luis quarries, it is very difficult to know which is the level that is actually present, because it is impossible to differentiate one from the other.

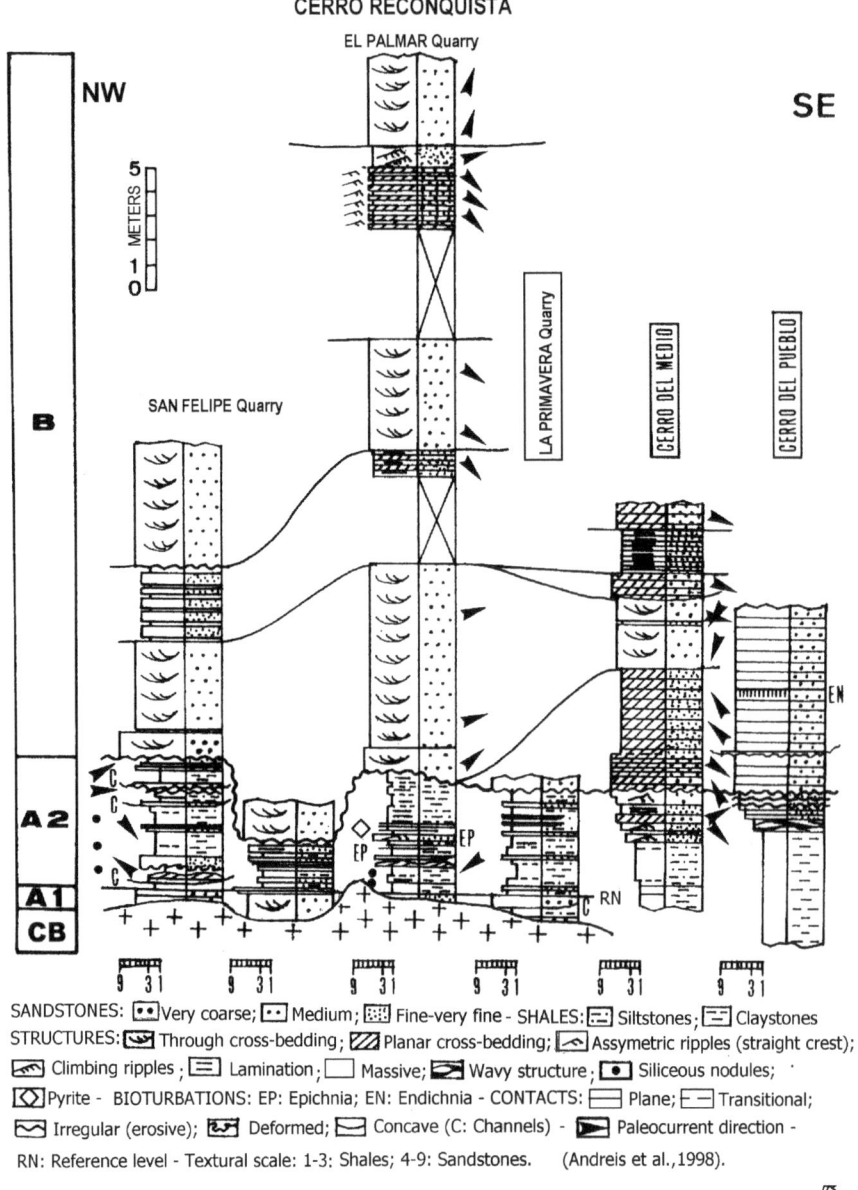

SANDSTONES: ⊡ Very coarse; ⊡ Medium; ⊞ Fine-very fine - SHALES: ⊟ Siltstones; ⊟ Claystones
STRUCTURES: ▦ Through cross-bedding; ▨ Planar cross-bedding; ⬛ Assymetric ripples (straight crest);
⬛ Climbing ripples ; ☰ Lamination; ⬜ Massive; ⬛ Wavy structure; ⬛ Siliceous nodules;
◇ Pyrite - BIOTURBATIONS: EP: Epichnia; EN: Endichnia - CONTACTS: ☰ Plane; ☰ Transitional;
⬳ Irregular (erosive); ⬲ Deformed; ⬱ Concave (C: Channels) - ▶ Paleocurrent direction -
RN: Reference level - Textural scale: 1-3: Shales; 4-9: Sandstones.     (Andreis et al.,1998).

**Fig. 4.3** San Manuel area. Proterozoic sedimentary sequence. La Juanita Formation: Cuarcitas inferiores (A1) and Miembro dolomías (A2), Cerro Largo Formation (B), crystalline basement (BC) Taken from Zalba and Andreis (1998)

The sequence continues upward with yellowish claystones and siltstones, with subordinate quartzite levels, also corresponding to the La Juanita Formation (or Villa Mónica Formation) with thicknesses of up to 8 m (Fig. 4.3: A2). Going upwards, separated by an erosion surface, the quartzites of the Cerro Largo Formation are found, (Fig. 4.3: B: Cuarcitas Superiores) with a total thickness of 33 m (Cerro Reconquista).

## 4.1.2 Residual Deposits

1. Los Cinco Nietos

Only the clays resulting from the alteration of the crystalline basement rocks were exploited in the past. The original basement rocks, reddish, purple, and white colored were hydrothermally argillized, and then altered by subaerial exposure (Zalba and Andreis 1998). These conclusions, based on field, petrographic and mineralogical studies, have been corroborated—after many years of discussion with other colleagues—through the study of rare earth elements in the basement rocks and in the overlying sedimentary sequence (c.f. Fernández et al. 2007).

In the stratigraphic section carried out at the Los Cinco Nietos quarry (Fig. 4.4: A) also we offer data on the composition of the clays and their impurities (Fig. 4.4: B) and the frequency of the phyllosilicates (Fig. 4.4: C). The composition of the phyllosilicates (determined by petrography and X-ray diffraction) is made up of pyrophyllite, kaolinite, muscovite, and illite with subordinate illite-smectite. Impurities are represented by abundant hematite and variable proportions of quartz, pyrite, clinozoisite, anatase, and goethite, according to the exploited level (see Fig. 4.4: B). Petrographic analyses indicate scarce quartz showing "caries" texture (Fig. 4.5 a) since the mineral has hydrothermally reacted with kaolinite to give pyrophyllite which grows with a cabbage-like texture (Fig. 4.5a, b). Figure 4.5c shows masses of hydrothermally formed pyrophyllite (Py) also replacing opaque minerals.

In the system $Al_2O_3-SiO_2-H_2O$ (Evans and Guggenheim 1988) pyrophyllite is stable over a narrow temperature range ($\sim 250$ to $\sim 350$ °C), at 1 and 2 Kbars pressure (Hemley et al. 1980) with the most frequent reaction producing pyrophyllite during prograde metamorphism through the reaction

$$Kaolinite + Quartz \rightarrow Pyrophyllite + Water.$$

## 4.1.3 Scanning Electron Microscopy of Residual Deposits: Los Cinco Nietos

SEM obtained from the saprock show a very compact texture of a weathered basement, where sheets of pyrophyllite, formed hydrothermally, are "locked" and "in situ" (Fig. 4.6a). Compare the texture of this hydrothermal pyrophyllite with the

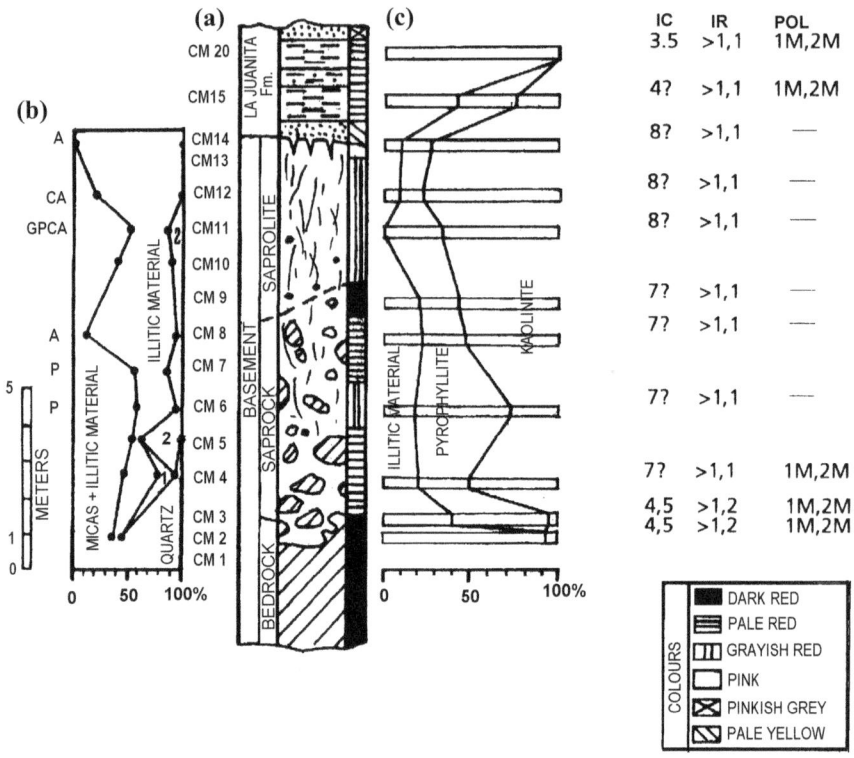

1- CABBAGES OF PYROPHYLLITE  - 2- QUARTZ REPLACED BY PYROPHYLLITE

A: ANATASE - C: CLINOZOOISITE - P: PYRITE - G: GOETHITE  - NR: REFERENCE LEVEL

IC: CRYSTALLINITY INDEX - IR: EXPANDABILITY INDEX -  POL: POLYTYPES.

**Fig. 4.4** San Manuel area, Los Cinco Nietos quarry. **a** Stratigraphic section showing the altered basement (bedrock, saprock and saprolite) and the Proterozoic sedimentary sequence. **b** Mineral composition of the basement rocks. **c** Percentage of phyllosilicates. Taken from Zalba and Andreis (1998)

one of detrital pyrophyllite, inherited from these basement rocks (Fig. 3.29, Chap. 3). Figure 4.6b represents "books" of mica, with open cleavage and partially deferrized, serrated edges produced by diagenetic dissolution. Diagenetic goethite can be seen in clustered aggregates (Fig. 4.6c).

2.  Julián Luis

Mineralogical analysis carried out in the in the upper part of the saprolite and sedimentary sequence superimposed in the Julián Luis quarry (Zalba and Andreis 1998) show the following results for Sample SMR (saprolite): illite + illite-smectite 15 %, pyrophyllite 46 %, kaolinite 39 %. Impurities include abundant quartz and feldspars, scarce iron oxides, and hydroxides (hematite and goethite).

**Fig. 4.5** Photomicrographs of San Manuel area. **a** Anhedral crystals of pyrophyllite (Py) and a few crystals of quartz (Qz) with "caries" texture. **b** Pyrophyllite aggregates which grows with a cabbage-like texture. (XPL, magnification x100). Taken from Zalba and Andreis (2001). **c** Masses of hydrothermally formed pyrophyllyte (Py) also replacing opaque minerales, saprock, San Manuel (XPL). Zalba et al. 2010

## 4.1.4  Sedimentary Deposits

1. Los Cinco Nietos

Clay sediments, which cover the residual deposits, lie between quartzite strata (La Juanita Formation). They are yellow and green to ocher colored. Petrographic studies (Zalba and Andreis 1998; Zalba and Andreis 2001) show an illitic and illite-smectite composition, with low participation of pyrophyllite and kaolinite. The impurities are abundant quartz and scarce feldspars, deferrized micas, and very little hematite and goethite. Pyrophyllite has been identified in intraclasts, derived from the erosion and transport of the already weathered underlying basement rocks (Fig. 4.7a). The presence of intraclasts of pyrophyllite proves that in the sediments it is detrital and not hydrothermal, as other authors sustain. These sediments are formed by clay material alternating with aligned quartz grains (Fig. 4.7b) and partially desferrized micas, also parallel to the stratification (Fig. 4.7c). Impurities are represented by abundant quartz and feldspars. These levels are not exploited. Figure 4.7d represents a very important document of MISS (Microbially Induced Sedimentary Structures) according to the definition of Noffke et al. (2001). Evidently, a very advanced step on diagenetic processes caused the fracture of

**Fig. 4.6** Scanning electron micrographs, Los Cinco Nietos quarry. Saprock: **a** Abundant sheets of pyrophyllite (Py) are "locked" and "in situ". **b** "Books" of mica (M), with open cleavage and serrated edges produced by diagenetic dissolution. Saprolite: **c** Goethite (G) crystals in clustered aggregates of diagenetic origin. Scale bar: 3 microns

hematized, dark mat relics (compare this picture with similar examples of MISS in the same lithostratgraphic unit: the Villa Mónica Formation, given previously for the Sierra La Juanita Sector: Chap. 3, Fig. 3.23a, b). Algal mat relics in the Villa Mónica Formation at San Manuel area are isolated and very difficult to recognize if one has not been able to detect other diagnostic associated features (e.g., cellular algal colonies) or to observe these structures in the context of biogenerated rock where original textures have not been totally erased, as in this case. Comparing diverse examples of MISS found in different localities of the Tandilia System we could finally understood the significance of these structures that, altogether reach their genuine significance. The final and most important channel link of this history is this last example shown in Fig. 4.7d.

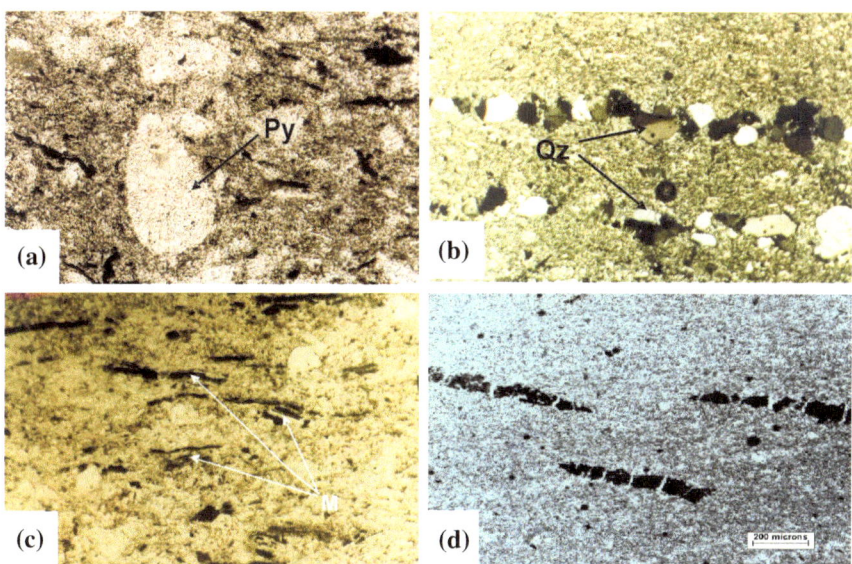

**Fig. 4.7** Photomicrographs of sedimentary levels of Los Cinco Nietos quarry. La Juanita Formation. **a** Pyrophyllite (Py) intraclasts in siltstones, derived from erosion of the basement rocks (PPL). **b** Bands of clay material alternating with aligned quartz (Qz) grains (XPL). **c** Partially deferrized micas (M), also parallel to the stratification, (XPL). Magnification 100x. **d** Note the similarity with Fig. 3.23b. Isolated and aligned hematized relics of fractured MISS in weathered stromatolitic dolostones completely replaced by illitic clays. The alignment is interpreted as representing original bedding planes. Villa Mónica Formation, San Manuel (PPL)

The finding of MISS in the Villa Mónica Formation, now at San Manuel deposits, corroborates the presence of biosignatures all over the Tandilia System, although diagenesis has almost erased any other signal of organic existence and the sediments are at present mainly composed of detrital illite and pyrophyllite, and diagenetic interstratified illite-smectite and kaolinite (Zalba and Andreis 1998).

2. Julián Luis

Sedimentary clay levels (La Juanita Formation) in the Julián Luis quarry show the following mineralogical composition (fraction < 2 microns) for sample SMC (sedimentary level), according to Garrido et al. (1998): illite 86 %, pyrophyllite traces, kaolinite 6 %, illite-smectite 6 %, smectite 6 %.

Sample SMC is characterized by the presence of abundant impurities of quartz, feldspars, and micas (detrital) and very scarce hematite and goethite (secondary).

Chemical analysis of major elements performed on samples of weathered basement rocks (residual red clay deposits) and on the sedimentary sequence of the La Juanita Formation (ocher potentially exploitable clay level) in the Julián Luis quarry can be checked in Table 4.1. Chemical analyses show that in the altered basement rock, the lower proportion of $SiO_2$ with respect to the sedimentary level is consistent with its reaction with kaolinite to give pyrophyllite by hydrothermal

**Table 4.1**  Chemical composition (weight percent) of residual red clay deposits (SMR) and sedimentary clays (SMC) of the La Juanita Formation, San Manuel area, Julián Luis quarry

| Sample | SiO$_2$ % | Al$_2$O$_3$ % | Fe$_2$O$_3$ % | MgO % | CaO % | Na$_2$O % | K$_2$O % | TiO$_2$ % | LOI % |
|--------|-----------|---------------|----------------|-------|-------|-----------|----------|-----------|-------|
| SMC | 62.01 | 21.79 | 3.44 | 0.90 | 0.21 | 0.02 | 6.89 | 1.34 | 4.09 |
| SMR | 39.92 | 29.09 | 18.00 | 0.13 | 0.17 | 0.63 | 1.57 | 2.25 | 6.76 |

Taken from Garrido et al. (1998)

processes that occurred in the original rock. The major amount of Al$_2$O$_3$ in residual deposits is due to a predominance of kaolinite and pyrophyllite in these rocks. Iron oxide comes from the weathering of micas in the altered basement rocks and appears as hematite and goethite, and its decrease in sedimentary levels is due to the concomitant reduction of hematite and goethite there.

TiO$_2$, abundant and originally contained in the micas, concentrate when the micas are destructed in the weathering profile. K$_2$O increase in the sedimentary levels is due to the abundance of illite-smectite and the presence of allochthonous, unaltered micas and illite in these deposits, as observed by optical microscopy. Calcium oxide is very scarce in both types of deposits (secondary calcite) and sodium oxide is greater in SMC levels, presumably contained in plagioclase, which is much more abundant in the sedimentary sequence.

### 4.1.5   Technological Properties of Residual (SMR) and Sedimentary (SMC) Clays of San Manuel, Julián Luis

Technological tests intended to establish correlations between the nature of the raw materials, the features of the final product and the conditions of processing, at laboratory scale. For identical treatment (preparation by moulding, thermal cycle, etc.) the product properties observed will depend mainly, on the mineralogical composition: clay minerals and mineral association, and on their grain size. The Pyrometric Cone Equivalent (PCE) values, particle size distribution and plasticity index are presented in Table 4.2. The greater refractoriness of SMR is due to the amount of pyrophyllite and the low content of alkalis. In terms of the index of

**Table 4.2**  Plasticity index, particle size distribution and Pyrometric Cone Equivalent (PCE) values of the San Manuel clays

| Technological test | SMR | SMC |
|--------------------|-----|-----|
| Plasticity index | 5.6 | 8.6 |
| Fraction < 44μm % | 35.2 | 74.3 |
| Fraction > 2 μm % | 9.4 | 16.3 |
| PCE, temperature (°C) | 1430 | 1260 |

Taken from Garrido et al. (1988)

plasticity values, higher in SMC clays, it is associated to illite-smectite content as well as a higher percentage of fine particles in those samples.

The results of heating tests according to the temperature are presented in Tables 4.3 and 4.4. Linear shrinkage, water absorption, apparent porosity, and mechanical resistance (MOR) of the obtained products are related to material differences. As one might expect, shrinkage, and density increase with the increase of temperature, while the porosity and water absorption decrease. The values of maximum contraction presented in Tables 4.3 and 4.4 are 8.3 % (SMR) and 13.8 % (SMC), corresponding the largest to the most plastic material. In the same way, the green resistance (MOR) is related to plasticity, since, generally, more plastic clays exhibit higher values of MOR. The mechanical strength of the specimens is low, although grows with the increase of temperature. A significant increase of the MOR at maximum temperature for sample SMR is probably associated with the formation of mullite, more than to the effect of porosity. In general, it has been observed that a dense microstructure, with low porosity and small pores due to cross-breeding of a continuous mullite phase crystals, improves the mechanical properties, resistance to abrasion and the hardness of the heated work piece.

**Table 4.3** SMC clay: Linear shrinkage (LS), water absorption (WA), apparent porosity (AA), and mechanical resistance (MOR) depending on the calcination temperature

| Temperature (°C) | Plastic mud state | | | | Pressed samples (200 kg/cm$^2$) | | | |
|---|---|---|---|---|---|---|---|---|
| | LS (%) | WA (%) | AP (%) | MOR (kg/cm$^2$) | LS (%) | WA (%) | AP (%) | MOR (kg/cm$^2$) |
| 100 | 0.2 | – | – | 4 | 0.2 | – | – | 2 |
| 950 | 0.8 | 19.7 | 33.9 | 77 | 0.2 | 14.2 | 27.4 | 87 |
| 1000 | 5.0 | 13.4 | 25.5 | 142 | 0.5 | 9.9 | 20.6 | 148 |
| 1050 | 8.6 | 6.5 | 14.3 | 197 | 3.8 | 4.5 | 10.5 | 205 |
| 1150 | 13.8 | 0.5 | 1.0 | 244 | 5.8 | 0.3 | 0.7 | – |
| Wáter % | 24 | | | | 5 | | | |

Taken from Garrido et al. (1998)

**Table 4.4** SMR clay: Linear shrinkage (LS), water absorption (WA), apparent porosity (AP), and mechanical resistance (MOR) depending on the calcination temperature

| Temperature (°C) | Plastic mud state | | | | Pressed samples (200 kg/cm$^2$) | | | |
|---|---|---|---|---|---|---|---|---|
| | LS (%) | WA (%) | AP (%) | MOR (kg/cm$^2$) | LS (%) | WA (%) | AP (%) | MOR (kg/cm$^2$) |
| 100 | 0.2 | — | — | — | 0.2 | — | — | — |
| 950 | 2.5 | 18.7 | 35.3 | 21 | −2.0 | 11.3 | 24.9 | 52 |
| 1000 | 3.7 | 17.6 | 34.0 | 53 | −2.0 | 9.4 | 21.6 | 119 |
| 1050 | 4.5 | 15.8 | 31.6 | 98 | -1.0 | 9.3 | 20.8 | 162 |
| 1150 | 8.3 | 10.7 | 23.2 | 208 | 1.6 | 5.7 | 13.0 | — |
| Wáter % | 20 | | | | 5 | | | |

Taken from Garrido et al. (1998)

Crystalline phase identification by X-ray diffraction indicates that the main reflections of quartz, feldspars, illite-smectite, and least amount of pyrophyllite persist at 950 °C and decrease gradually with increasing temperature. Mullite, cristobalite, and hematite were identified in the SMR sample treated at 1100 °C. This result is consistent with the reaction products that are developed from pyrophyllite, forming mullite with abundant silica release. In the X-ray pattern of sample SMC heated at 1150 °C, quartz is recognized. Based on the analysis by X-ray diffraction it is not possible to recognize other crystalline species at this temperature, due to the low intensity of the reflections. In the current literature, there is still disagreement on the crystalline phases in which decomposes the illitic material at these temperatures.

Results from Table 4.3 indicate that, with the SMC sample, dense, and slightly colored products are obtained. By heating the sample at 1150 °C, water absorption decreases to values less than 5 %, with regular contraction. These materials could be used as a partial substitute for other clays or fluxes to increase the speed of sintering of ceramic masses for various applications (crockery, sanitary, tiles, etc.). According to obtained results for SMR sample (Table 4.4) this clay presents lower shrinkage and higher porosity in the heated product, even at maximum temperature of 1150 ° C. These values are probably linked to the presence of pyrophyllite and greater grain size that characterize this sample. However, the low plasticity and green resistance of this material must be taken into account to determine its possible use in the industry. According to the existing literature, this problem could be solved, in part, by the addition of a small percentage of plasticizers. The quality of the material makes it suitable for its use in blends and to correct excessive plasticity and contraction in red ceramic formulations as well as to improve porosity (Garrido et al. 1998). Therefore, the presence of pyrophyllite in residual clays increases their refractoriness, but they contain between 15 and 25 % of iron oxides and hydroxides. Technological tests indicate that residual materials are of regular plasticity and are suitable for the manufacture of red and structural ceramics (Schalamuk et al. 1992).

Sedimentary clays, because of their high quartz content, the illite and mica predominance, their low water absorption and light colors after charred, are suitable for the manufacture of clear ceramics and faience. According to Schalamuk et al. (1992) other uses for these sedimentary clays are found as loading material (rubber), when they are of easy dispersion, and as fluxes, when the content of alkali is high. Those with high content of iron oxides could also be used in the preparation of common bricks.

# References

Dristas JA, Frisicale MC (1984) Estudio del yacimiento de arcillas del Cerro Reconquista, San Manuel, Sierras Septentrionales de Buenos Aires. In: Proc 9° Congreso Geológico Argentino, Actas, Buenos Aires, 5:507–521

Evans BW, Guggenheim S (1988) Talc, pyrophyllite, and related Minerals. In: Bailey SW (ed) Hydrous Phyllosilicates (exclusive of micas). Rev Mineral 19(8):225–280

Fernández R, Tessone M, Etcheverry R, Caballé M, Coriale N, Echeveste H (2007) Distribuciones de elementos de las Tierras Raras en el basamento alterado de Tandilia: zona de San Manuel. In: Proc 6° Jornadas Geológicas y Geofísicas Bonaerenses, Mar del Plata, Abstract, p 40

Frisicale MC (1991) Estudio de algunos yacimientos de arcilla originados por actividad hidrotermal en las Sierras Septentrionales de la Provincia de Buenos Aires. Tesis Doctoral, (inédito). Universidad Nacional del Sur, Bahía Blanca, 217 pp

Garrido L, Zalba PE, Morosi M (1998) Estudio tecnológico de arcillas de la zona de San Manuel, Buenos Aires, Argentina. 9° Congreso Internacional de Cerámica y 9° del MERCOSUR, Olavarría. Revista Cerámica y Cristal, pp 53–54

Hemley JJ, Montoya JW, Marinenko JW, Luce RW (1980) Equilibria in the system $Al_2O_3$–$SiO_2$–$H_2O$ and some general implications for alteration/mineralization processes. Econ Geol 75: 210–228

Noffke N, Gerdes G, Klenke T, Krumbein WE (2001) Microbially induced sedimentary structures —a new category within the classification of primary structures. J Sedimentary Res 71:649–656

Schalamuk I, Etcheverry R, Garrido L, Fernández R (1992) Geología y características tecnológicas de los depósitos de arcilla de los partidos de Azul y Lobería, provincia de Buenos Aires. In: Proc 4° Congreso Nacional y 1° Congreso Latinoamericano de Geología Económica, pp 477–488

Zalba PE (1988) Arcillas de las Sierras Septentrionales de Buenos Aires. Publicación Especial N° 1, CETMIC-CIC, Provincia Buenos Aires, La Plata, 62 p

Zalba PE, Andreis RR (1998) Basamento cristalino saprolitizado y secuencia sedimentaria suprayacente en San Manuel, Lobería, Sierras Septentrionales de Buenos Aires, Argentina. In: Proc 7° Reunión Argentina de Sedimentología, Salta, pp 143–153

Zalba PE, Andreis RR (2001) Stratigraphy, sedimentology and mineralogy of Neoproterozoic clay deposits, Sierras de Tandilia, Province of Buenos Aires, Argentina. Economical Importance. 12th International Clay Conference, Pre-Simposium Field Trip, Bahía Blanca, 80 p

Zalba PE, Morosi ME, Manassero M, Conconi, MS (2010) Microscale diagnostic diagenetic features in Neoproterozoic and Ordovician units, Tandilia basin, Argentina: a review. J Appl Sci 10(22): 2754–2772

# Chapter 5
# Olavarría County

**Abstract** In the Sierras Bayas Sector clay deposits correspond to the Olavarría Formation and to the Cerro Negro Formation, both of Neoproterozoic age. They represent important sedimentary deposits of the Tandilia System because of their extension and technological applications. The Cerro Largo Formation is spread along the Sierras Bayas area and it consists of three units separated by transitional contacts. Geologic-stratigraphical detail studies led to the redefinition of the unit 3 of the Cerro Largo Formation considering it as a separate unit and named Olavarría Formation, studied in various localities of the Sierras Bayas. The unit is limited by erosive unconformities that separate it from the underlying unit (Cuarcitas Superiores) and the overlying one (Loma Negra Formation). MISS, confirm the presence of algal mats also in the Olavarría Formation. X-ray diffraction, optical, microscopy and SEM have been performed on the clay deposits. At La Pampita quarry, the finding and study of a mud bed—connected to several mud pipes—intercalated in the Loma Negra Formation limestones—are attributed to upward mud injection from the underlying Olavarría Formation (telogenesis). In general, the clays have low plasticity index. They have low-refractoriness, and are classified as varied red clays and varied ocher clays. They are used in red ceramics. The Cerro Negro Formation has been found in the subsoil and by the opening of several quarries near the Cerro Negro village. The Cerro Negro Formation was divided in two depositional systems separated by an irregular paleosuperface: the Lower Depositional System and the Upper Depositional System. The mineralogical composition of the clays is different in the two depositional systems. Technological tests have been carried out in "composite" samples (stocks) for each of the depositional systems. The clays of the two depositional systems present similar Pyrometric Cone Equivalent (PCE). They are exploited in the area of Cerro Negro, with excellent results and large reserves, for their use in the manufacture of ceramic floor coverings.

**Keywords** Sierras Bayas · Geology · Stratigraphy · Mineralogy · Sedimentary clays · Olavarría Formation · Cerro Negro Formation · MISS · Mud bed · Mud pipes · Telogenesis · Technology

© Springer International Publishing Switzerland 2016         107
P.E. Zalba et al., *Gondwana Industrial Clays*,
Springer Earth System Sciences, DOI 10.1007/978-3-319-39457-2_5

## 5.1   Sierras Bayas Sector

### 5.1.1   Sedimentay Deposits; Characteristics, Mineralogical and Chemical Composition of the Olavarría Formation

Clay deposits in this sector correspond to the Olavarría Formation (for other authors included in the Cerro Largo Formation) and to the Cerro Negro Formation (see Table 1.1), both of Neoproterozoic age. They represent important sedimentary deposits of the Tandilia System because of their extension and technological applications. The Cerro Largo Formation is spread along the Sierras Bayas area and was described by Poiré (1987, 1993) as consisting of three units (or members) separated by transitional contacts. From bottom to top:

- Interbedded claystones and sandstones (including siliceous breccia, mudstones, and diamictites)
- Quartzites (known as Cuarcitas Superiores)
- Claystones and siltstones.

On the basis of geologic-stratigraphical detail studies, Andreis et al. (1996) redefined the Claystone Member (unit 3) of the originally defined Cerro Largo Formation, considering it as a separate unit and named Olavarría Formation, which was studied by the authors of the present work in various localities of the Sierras Bayas: first, at San Alfredo quarry, Olavarría, where the type area was defined: El Tajo. Also, the Olavarría Formation was studied at Canteras Fiscales, Cal Moreno, Aust, San Andrés, Cementos Avellaneda, Minera Olavarría, LOSA, and more recently (Zalba et al. 2007) at the La Pampita quarry, Loma Negra, among others. In Fig. 5.1, from base to roof, different colors (gray, ocher, and reddish) of the clay deposits of this formation can be seen at the Canteras Fiscales quarry. In this work the name of Olavarría Formation is adopted for unit 3 of the Cerro Largo Formation of Poiré (1987). It is limited by erosive unconformities that separate it from the underlying unit (Cuarcitas Superiores) and the overlying one (Loma Negra Formation). In

**Fig. 5.1** Olavarría Formation, Canteras Fiscales quarry, Sierras Bayas. Gray clays (*1*), ocher clays (*2*), reddish clays (*3*). Overlying and unconformably, lies the Loma Negra Formation

**Fig. 5.2** Olavarría
Formation, Los Hermanos
quarry, Sierras Bayas. Ocher
clays (*2*), reddish clays (*3*),
unconformably covered by
the Loma Negra Formation.
The *arrows* indicate the
relative movement of the
blocks

Fig. 5.2 red claystones of the Olavarría Formation can be seen in contact with the
Loma Negra Formation limestone, which has descended by faulting. The Olavarría
Formation includes clayey, hetherolitic, and sandy facies (quartzites) with thickness
of up to 25 m (38 m in the subsoil).

The Olavarría Formation, at Sierras Bayas (stratigraphic section El Tajo type
area, Fig. 5.3) is made up of three sections: a lower section, with more than 11 m
thick reddish, gray, yellowish, and ocher claystones with interbedded siltstones, and
white quartzites. A middle section, up to 9 m thick, consists of alternating sandstones and claystones, as well as claystone levels, beige, and ocher colored. An
upper section (5 m thick) is composed of siltstones and subordinate, reddish and
lilac claystones which predominate over yellowish or grayish colors. All are economically productive.

1. Canteras Fiscales
   Petrographic studies on clays from Canteras Fiscales quarry (Zalba et al. 1996)
   Olavarría Formation, allowed the authors to identify, in the gray clays, a texture
   where lamination is formed by the alternation of clay material and quartz. When
   the clay material predominates, it is colored with iron oxides and sulfides (hematite and pyrite, respectively: (Fig. 5.4a, b). The texture of the ocher clays
   offers a more pronounced lamination and iron oxide disposes preferentially on
   clays and along cracks and fissures (Fig. 5.4c). In the red siltstones (Fig. 5.4d),
   on the other hand, besides a good lamination and iron oxides along transversal
   cracks, secondary calcite filled cavities parallel to the lamination. But the most
   conspicuous sign is the presence of MISS, exemplified by very thin iron oxide
   bands developed in siliciclastic (illitic) sediments disposed parallel to the lamination, confirming the presence of algal mats also in the Olavarría Formation.
   Figure 5.5a, b are from the same sedimentary deposits (Olavarría Formation,
   Canteras Fiscales quarry, Sierras Bayas) and illustrate very thin microbial mats
   developed in siliciclastic (illitic) sediments, also recognized as MISS. The
   curved, dome, dark surfaces are considered to be bedding planes and the
   geometry of the domes do not superpose with each other upwards, that is why
   the same colony could not have grown after a renewal of the sediment supply
   (Fig. 5.5b).

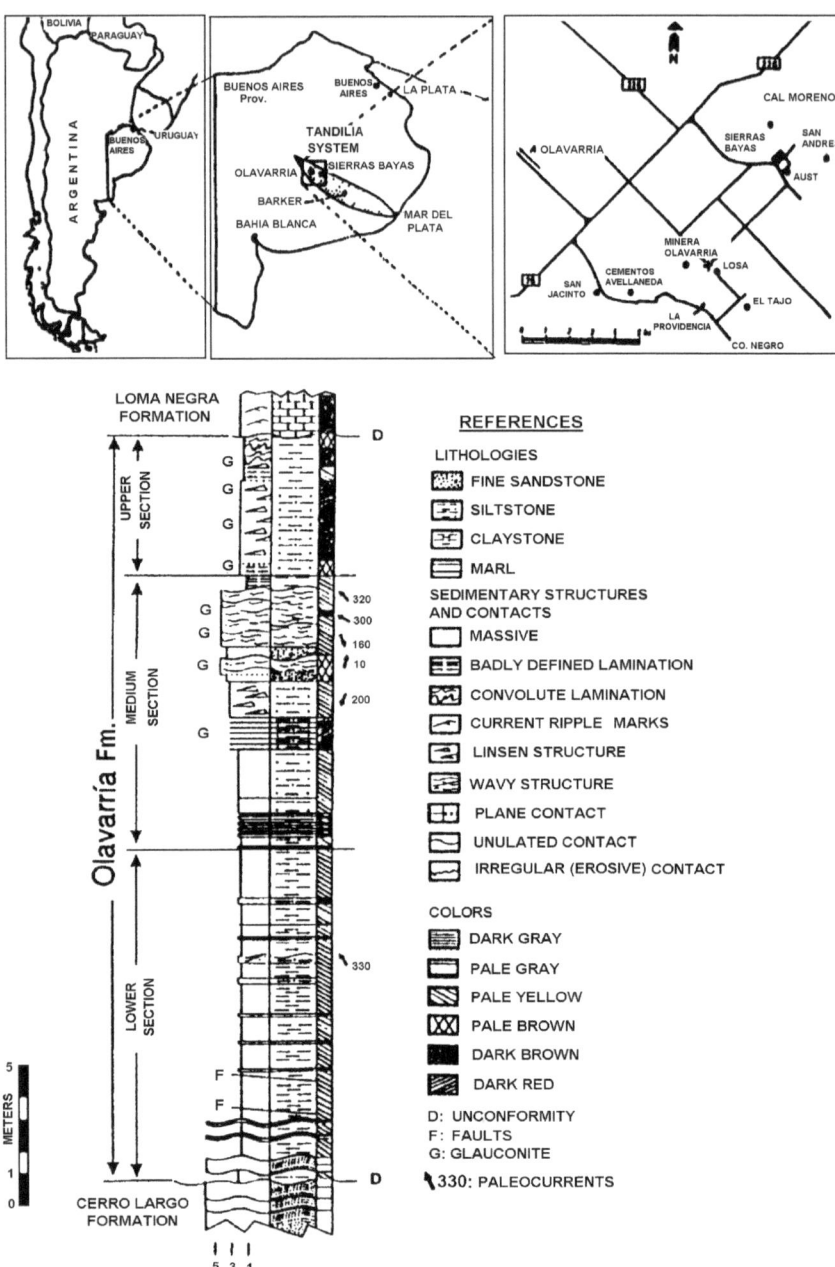

**Fig. 5.3** Location map and stratigraphic section of El Tajo: type area of the Olavarría Formation, Sierras Bayas. Taken from Andreis et al. (1996)

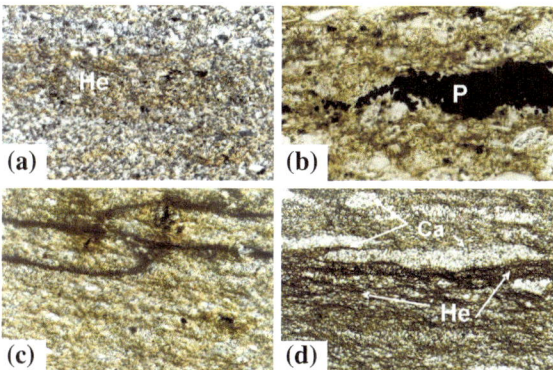

**Fig. 5.4** Petrographic study of the Olavarría Formation, Canteras Fiscales quarry. *Gray clays* **a** Clay material colored by hematite (He), (XPL, magnification 100×). **b** Clay material colored by pyrite (P), (XPL, magnification 400×). *Ocher clays* **c** Hematite filling cracks, (XPL, magnification 100×). *Reddish siltstone* **d** Diagenetic calcite (Ca) and Hematite (He) replacing algal mats (MISS), XPL, magnification 40×)

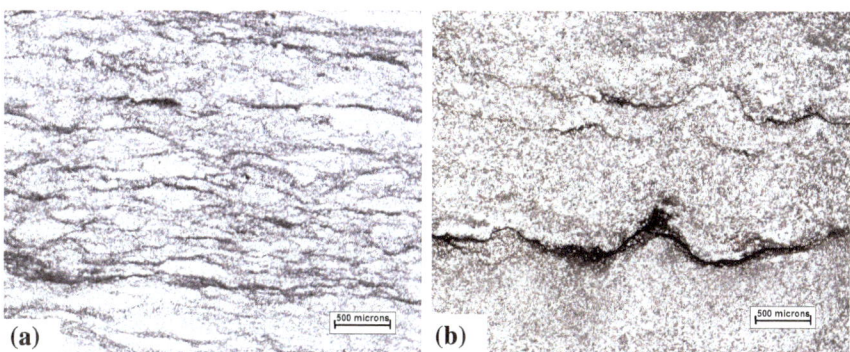

**Fig. 5.5** Photomicrographs of the Olavarría Formation, Canteras Fiscales quarry. MISS in Carbonate/Silicicalsic facies: **a** Dark, hematized, crinckly lamination as relics of microbial activity (PPL). **b** MISS where domes do not superpose upwards, separated by the renewal of continental input between microbial cycles (PPL). Taken from Zalba et al. (2010)

2. El Tajo Section, Loma Negra

   X-ray diffraction of the Olavarría Formation clays (Table 5.1) shows that they are composed of illite (I) + illite-smectite (I/S) which are predominant in the three sections, with similar percentages. Traces of calcite, abundant micas, and scarce feldspars are also present in the upper section. Quartz is very abundant, in all sections. Detrital micas, quartz, feldspars, and illite derive from the erosion of crystalline basement rocks while the I/S are the product of superimposed diagenetic processes. Scarce kaolinite and glauconite have been detected in the middle section, while traces, or very scarce smectite, is observed in the lower

**Table 5.1** Mineralogical composition by X-ray diffraction of clays of the El Tajo section, Sierras Bayas, the Olavarría Formation

| Sample | Illitic material (%) | Kaolinite (%) | Smectite (%) | Impurities | | |
|--------|---------------------|---------------|--------------|------------|---------|---------|
| | | | | Quartz | Feldspar | Calcite |
| T28 | 100 | – | – | v.a | – | v.a |
| T27 | 100 | – | – | v.a | – | v.a |
| T26 | 100 | – | – | v.a | v.s | v.a |
| T25 | 100 | Loma Negra Fm. | – | | v.s | |
| T24 | 100 | Olavarría Fm. | – | v.a | v.s | – |
| T23 | 100 | – | – | v.a | v.s | – |
| T22 | 100 | – | – | v.a | v.s | – |
| T20 | 100 | 4 | – | v.a | v.s | v.s |
| T18 | 100 | – | – | v.a | v.s | v.s |
| T17 | 96 | 8 | – | v.a | s | – |
| T16 | 100 | 6 | traces | v.a | v.s | – |
| T15 | 100 | 6 | traces | v.a | v.s | – |
| T13 | 92 | 20 | – | v.a | – | v.s |
| T12 | 94 | – | – | v.a | – | s |
| T11 | 94 | 5 | – | v.a | – | v.a |
| T9 | 80 | – | – | v.a | – | v.a |
| T6 | 100 | – | – | a | – | s |
| T4 | 100 | – | – | v.a | – | s |
| T3* | 100 | – | – | v.a | s | s |
| T2 | 100 | – | – | v.a | – | – |

*Presence of siderite
*References* v.a very abundant, *a* abundant, *s* scarce, *v.s* very scarce. *T0–28* samples
Taken from Zalba et al. (1996)

and middle sections. Iron oxides and hydroxides (hematite and goethite, respectively) are abundant in red clays while abundant pyrite, without altering, or altered to hematite, was detected in the lower section (gray clays). All these minerals are diagenetic.

Chemical analyses for selected samples of the stratigraphic profile carried on El Tajo, are shown in Table 5.2. Chemical analyses of major elements on ocher and red "stock pile" (or composite samples) of the Olavarría Formation, in the Sierras Bayas (Garrido et al. 1996) are shown in Table 5.3. "Stock pile" chemical data reflect differences between both materials. The contents of $K_2O$,

**Table 5.2** Chemical composition of selected samples of the El Tajo section, Sierras Bayas

| Sample | $SiO_2$ (%) | $Al_2O_3$ (%) | $Fe_2O_3$ (%) | MnO (%) | MgO (%) | CaO (%) | $Na_2O$ (%) | $K_2O$ (%) | $TiO_2$ (%) | $P_2O_5$ (%) | LOI (%) |
|--------|-------------|---------------|---------------|---------|---------|---------|-------------|------------|-------------|--------------|---------|
| T3 | 66.90 | 16.47 | 3.61 | <0.01 | 1.14 | 0.35 | 0.06 | 5.36 | 0.90 | 0.08 | 4.79 |
| T7 | 73.19 | 12.67 | 4.19 | 0.02 | 0.70 | 0.64 | 0.04 | 3.81 | 0.68 | 0.06 | 4.17 |
| T8 | 67.94 | 16.44 | 2.74 | <0.01 | 0.99 | 0.45 | 0.06 | 4.84 | 0.90 | 0.06 | 4.83 |
| T16 | 66.50 | 16.29 | 3.39 | <0.01 | 1.24 | 0.40 | 0.05 | 5.23 | 0.80 | 0.07 | 4.98 |
| T18 | 63.65 | 15.37 | 7.11 | <0.01 | 1.39 | 0.36 | 0.06 | 5.26 | 0.73 | 0.11 | 5.00 |
| T20 | 67.89 | 12.04 | 7.21 | <0.01 | 1.33 | 1.44 | 0.06 | 4.22 | 0.61 | 0.99 | 4.12 |

Taken from Zalba et al. (1996)

**Table 5.3** Chemical composition of ocher and red "stock pile" (or composite samples) of the Olavarría Formation

| Sample | SiO$_2$ (%) | Al$_2$O$_3$ (%) | Fe$_2$O$_3$ (%) | MgO (%) | CaO (%) | Na$_2$O (%) | K$_2$O (%) | TiO$_2$ (%) | LOI (%) |
|--------|-------------|------------------|------------------|---------|---------|-------------|------------|-------------|---------|
| Red | 64.4 | 15.8 | 7.04 | 1.33 | 0.31 | 0.06 | 5.74 | 0.79 | 4.48 |
| Ocher | 67.6 | 13.14 | 4.5 | 0.87 | 1.48 | 0.06 | 4.20 | 0.77 | 5.09 |

Taken from Garrido et al. (1996)

Fe$_2$O$_3$, and the Al$_2$O$_3$/SiO$_2$ relationship are greater in the red material with respect to the ocher one. In the latter, an increase in CaO values is recorded on account of the presence of calcite. The proportion of K$_2$O in both cases is less than 10 % and indicates the presence of expansive layers in the illite, while the amount of MgO is consistent with the proportion of this oxide contained in illites derived from the weathering of feldspar of granitic rocks. The relationship Al$_2$O$_3$/K$_2$O is relatively higher in the ocher clay (close to 3). Calculations made on separated stock piles of ocher and red clays (composite sample) reach 37 % illite in the ocher material and 50 % illite for the red material, respectively. The amount of silica includes quartz (free silica) and is close to 50 % for the ocher material, and 40 % for the red one, respectively.

3. La Pampita, Loma Negra

   Studies carried out in a quarry front opened in the vicinity of the cement plant L´ Amalí, Loma Negra (Fig. 5.6), located 14 km to the SW of the city of Olavarría, deserve a special mention in the Sierras Bayas Sector because of the finding of a mud bed—connected to several mud pipes—intercalated in the Loma Negra Formation limestones, and whose study helped the authors in the interpretation of complex processes of diagenesis occurred regionally, related the Olavarría Formation and the Las Aguilas Formation (Cuchilla de Las Aguilas locality)

**Fig. 5.6** View of the opening of the La Pampita quarry, Loma Negra, Sierras Bayas. Arrows indicate the north–south front, where the study was carried out. A Clays of the Olavarría Formation. C Limestones of the Loma Negra Formation. The image was captured using the publicly available software Google Earth (http://earth.google.com/). Copyright 2008 DigitalGlobe; Europa Technologies

linked to the uplift of the Ventania System, Buenos Aires province (located 150 km SW from the Tandilia System) during the Middle Permian (Zalba et al. 2007). The concepts developed in that piece of work related to the Olavarría and the Loma Negra formations will be commented in the following pages.

Deformational structures have been described in the claystones of the Olavarría Formation in the Sierras Bayas area, by previous authors. Massabie et al. (1992) documented mud diapirs and attributed them to tectonic shearing of presumed Devonian to Permian-Triassic age. Sellés Martínez (1994) also described clay diapirs and hydraulic breccias in the Loma Negra Formation in the same area and suggested that these structures could be related to the diagenetic history of the Tandilia System. Also, Andreis and Zalba (1989) maintained that folding and other structural features found in the Tandilia System (Las Aguilas Formation) give evidence for compressive movements from the SW and associated them with the deformation of the Ventania System during Gondwana times. The nearby and most probable deformational phase that may have caused injection structures and induced important mineralogical changes is the folding that affected the Ventania System. A well-documented Permian age for this event is based on Rb–Sr and K–Ar geochronological data, mostly from white mica and illite of the sediments (Von Gosen and Buggisch 1989).

## 5.1.2   Discussion of Mud Beds and Pipes in the Sierras Bayas Area

A subhorizontal, north-south-oriented mud bed is exposed along 300 m at the La Pampita Quarry (14.5 km SW of Olavarría city, Table 1.1) within a 15 m thick limestone outcrop of the Loma Negra Formation, Sierras Bayas area (Fig. 5.7) and has been studied by Zalba et al. 2007. The mud bed, with concave-up planes, shows a nearly consistent thickness of 1.50 m, with localized gaps of about 0.20 m wide

**Fig. 5.7** General view of a north–south limestone front at La Pampita quarry, near Olavarría, Sierras Bayas area. The *vertical arrows* indicate a subhorizontal mud bed within the limestones of the Loma Negra Formation. The *horizontal arrows* indicate a mud pipe, connected with the mud bed. Taken from Zalba et al. (2007)

(Fig. 5.8a, b), and it overlies a lenticular polymictic orthoconglomerate of 0.05–0.
20 m thick. Both the lower and the upper contacts between the mud bed and the
limestone are parallel to the local bedding. Dissolution features of the limestone are
obvious at the contact with the clay material. The structures of the limestones vary
according to their location compared to the mud bed. Underlying limestones are
massive and rather undisrupted, whereas overlying ones are dislocated and displaced along vertical open fractures. Several small grabens have been observed in
the overlying limestones, with downward collapsing blocks (up to 3 m long)
leading to thinning of the mud bed (Fig. 5.8b). A NE–SW fault system has been
observed in the quarry.

Clasts and rock fragments (from millimeters to 5 cm long) of the polymictic
orthoconglomerate are irregular in shape, ranging from subangular to well-rounded
and showing a predominantly random orientation. Nevertheless, some large,
elongate fragments show a preferred N–S alignment of their longest axis. The
matrix is composed of very scarce, fine to coarse sand grains cemented by calcite.

Several vertical mud pipes have been observed in the La Pampita quarry, like the
one shown in Fig. 5.7. All of them are connected to the aforementioned mud bed
and intersect the bedding of the overlying limestone over distances which vary from
2 m up to the present erosion surface (7 m). It should be noted that downward mud
pipes have never been observed in the underlying limestones.

Small blocks of limestone were found in the clay pipes. The contacts between
the clay pipes and the host rock (limestone) exhibit the following features:
(1) marked deformation of disrupted limestone strata with a similar subvertical
reorientation on both sides of the clay pipe (Figs. 5.7 and 5.8a), and (2) consequent
dissolution of the disrupted limestone strata. At least six other clay pipes of the
same style (Fig. 5.8c) are exposed along the contact between the Loma Negra
Formation and the underlying clay deposits of the Olavarría Formation, 200 m NE
from the first limestone front described, and also in the Cementos Avellaneda
quarry, 3.2 km NE from the La Pampita quarry. All of these other mud pipes are
directly connected with the Olavarría Formation. The erosional unconformity
between the Olavarría and the Loma Negra Formations (Andreis et al. 1996) has
been partially obliterated by the development of a continuous dissolution front of
the overlying limestone (Fig. 5.8c). Several features related to compressive effects
and folding of the strata have been observed in the underlying Olavarría Formation.
The polymictic orthoconglomerate fabric exhibits chemical-alteration features:
corroded quartz grains with dissolution texture, replacement of quartz, and feldspars
by calcite, dissolution of feldspar crystals, and intergranular stylolites parallel to
bedding. The altered debris is cemented by calcite, with sparitic, spherulitic, aci-
cular, granular, or poikilotopic textures. Grain-breaking and displacement of quartz,
microcline, quartzite, and clay fragments are here attributed to fluid pressure producing a floating texture (Fig. 5.8a). Petrographic studies show that the sparite
records up to three successive growing pulses, displacing grains, replacing the
matrix, and filling cracks and fissures (Fig. 5.9a, b, c, d). The contact between the
conglomerate and the underlying limestone shows dissolution-precipitation features
leading to chalcedony and calcite crystallization in voids. Late mineralization
events comprise hematite precipitation in cement void space (Fig. 5.9b, d).

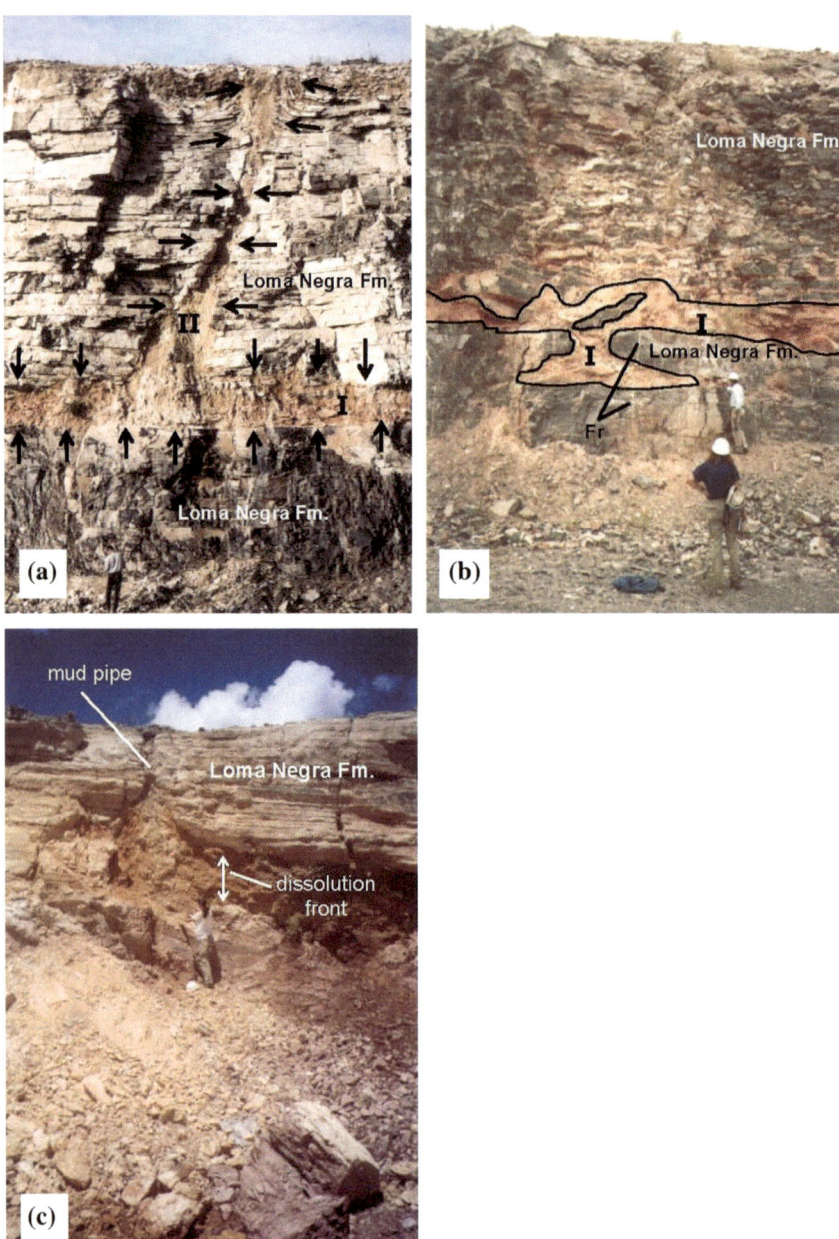

**Fig. 5.8** La Pampita quarry, Sierras Bayas area. **a** *I* Subhorizontal mud bed deposit within the Loma Negra Formation (vertical arrows). *II* Mud pipe associated with the mud bed (*horizontal arrows*). Upward deformation of the limestone strata at both sides of the clay pipe (*inclined arrows*). **b** Breaking and displacement of limestone blocks within the mud bed. Vertical fractures (Fr) infilled with calcite show only vertical displacement. **c** Continuous dissolution front developed at the base of the Loma Negra Formation. Mud pipe within the Loma Negra Formation. Taken from Zalba et al. (2007)

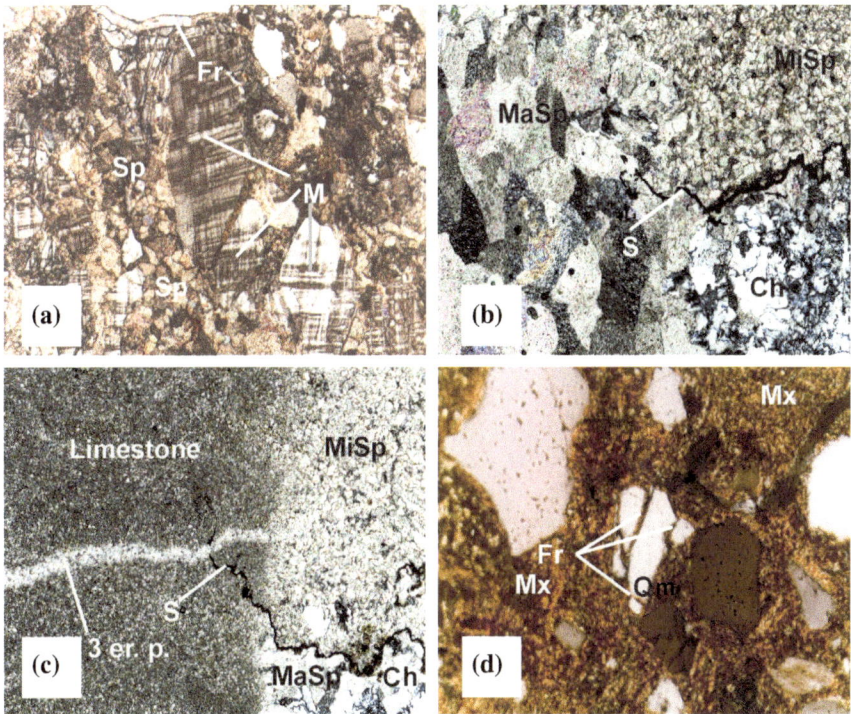

**Fig. 5.9** Photomicrographs of La Pampita quarry, Loma Negra Formation, Sierras Bayas area. **a** Polymictic orthoconglomerate. Microcline crystaloclast (M) encircled by coarse and fine sparite cement (Sp). Fractures are infilled by sparitic cement (Fr). **b** Detail of microsparite and macrosparite pulses (MiSp and MaSp, respectively) and chalcedony (Ch) void infilling. Stylolites (S) are cut by the growing of macrosparite cement. **c** Close-up of the third pulse (3rd. p.) of sparite cement filling fractures and cutting the stylolites (S) at the contact between the limestone and the conglomerate. **d** Thin section of the mud bed. Monocrystalline quartz (Qm) fractured (Fr) and displaced by injection matrix (Mx) of I/S composition. XPL, magnification ×100. Taken from Zalba et al. (2007)

The stages of the development of cements in the conglomerate and in the contact between the limestone and the conglomerate can be summarized as follows:

- Development of microsparite cement filling all the pore space available (first pulse).
- Recrystallization of intergranular microsparite into macrosparite (second pulse) arranged in a poikilotopic texture.
- Fracturing of the framework and the cements and filling of the cracks with a third pulse of sparite cement.
- Opal precipitation in voids and recrystallization to chalcedony usually restricted to the cement contacts and dissolving macrosparite and microsparite cements.
- Crystallization of hematite around clasts as well as in cement voids.

The overlying mud bed consists of clasts and small blocks (rock fragments ranging from millimeters to 3 cm) supported by a clay matrix. Rounded quartz (monocrystalline) and angular feldspar (microcline, orthoclase, and plagioclase) crystals predominate over lithic grains (elongate clay fragments, rounded granite–cataclasite, orthoquartzite, chert, and polycrystalline quartz). The sediments are poorly sorted and texturally immature, with features of grain fragmentation (Fig. 5.9d). The light brown clay matrix shows slickensides, and preferred orientation of clay minerals can be seen in the matrix which fills space between the disrupted grains. Tangential clay and hematite coatings are present around grains and as fracture-filling material. Aggregates of coarse-grained calcite (up to 0.50 m long) and rounded to elongated barite clasts (0.5–2 mm long) are found. The sediments are poorly lithified (no cementation features) and without any stratification or grading.

X-ray diffraction patterns from clay material from the subhorizontal mud bed at La Pampita quarry (Fig. 5.10a diffractogram) indicate that random I/S (with 70 % smectite) is the predominant clay-mineral component. A similar type of random I/S (up to 90 % smectite) has also been identified in the mud pipes connected with the Olavarría Formation (Fig. 5.10b diffractogram) related to the dissolution front developed at the contact between the Olavarría Formation and the Loma Negra Formation. At the base of the Olavarría Formation clay minerals consist of illite and ordered I/S (with <15 % smectite, Fig. 5.10c diffractogram) similar to those found regionally in this unit (Zalba et al. 1996).

### 5.1.3  Origin and Mechanisms of Formation of Mud Beds and Pipes

The orthoconglomerate and the mud bed and pipes contain rock fragments from:S

- Crystalline basement rocks: orthoclase, microcline, polycrystalline quartz, and granite-cataclasite grains
- Cerro Largo Formation: orthoquartzite fragments

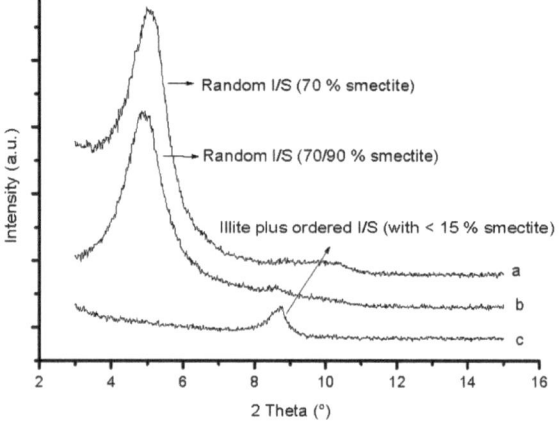

**Fig. 5.10**  Mineralogical composition of clay minerals by X-ray diffraction. La Pampita quarry, Sierras Bayas area. Mud bed: *a* Random I/S (70 % smectite). Mud pipe directly connected with the Olavarría Formation: *b* Random I/S (90 % smectite). Base of the Olavarriia Formation: *c* Illite + ordered I/S (with, 15 % smectite). Taken from Zalba et al. (2007)

- Olavarría Formation: illitic clays
- Loma Negra Formation: chert and limestone fragments

Both the mineralogical and the textural characteristics of the clay material observed within the poorly lithified mud bed and pipes intersecting the limestones of the Loma Negra Formation are consistent with the features of mud injection, which can occur in response to penetration of over pressured, fluidized muds along previously developed faults and fractures and/or sedimentary discontinuities within the limestones. The regular thickness of the mud infilling, the dissolution of the host limestone, and the precipitation of secondary minerals in open cracks and pores as well as the grain fragmentation are also typical of mud injections, as stated by Pickering et al. (1988). The preferential orientation of the clay particles in the matrix which fills the spaces between the disrupted grains and the tangential clay coatings observed around the grains are consistent with a process of mechanical infiltration. Additionally, the slickensides in the massive clay material are indicative of the rearrangement of the clay particles due to the expulsion of the solutions and the extrusion of the highly viscous remaining clay paste (Higgins and Saunders 1967).

The same structural style of vertical mud pipes found in different levels of the Loma Negra Formation (existence of cone-like structure within the adjacent limestone), the direct connection between some mud pipes and the underlying Olavarría Formation, and the dissolution features at the wall of the host limestone indicate that the upward mud injection originated in fluidization and transport of mud material from the underlying Olavarríia Formation. It is also likely that mud injection was associated with strong fluid–rock interaction which resulted in the dissolution of calcite (by acidic fluids) and the crystallization of a random I/S (with 70–90 % smectite) at the expense of the ordered I/S (with <15 % smectite) previously formed during the peak diagenesis of the underlying Olavarría Formation. The processes described herein do not refer to weathering but to alteration by injection of over pressured fluids whose acidic properties originated from the presence of maturating organic matter in the source formation (the underlying Olavarría Formation).

Accumulation of over pressurized acidic compaction waters within mudstones located below a seal caprock (such as impermeable limestone) is a common feature during the burial diagenesis of sedimentary clay formations (cf. Burley and MacQuaker 1992; Worden and Burley 2003), and such a phenomenon can persist as long as the pore pressure at the top of the mudstone does not exceed the mechanical resistance of the seal. Basically, acidic fluids are released by maturing organic-rich shales during burial diagenesis (mesogenesis) and acidic pore solutions leach carbonate cement and grains. Also, connate waters expelled during diagenetic processes are well known to be acidic and rather reducing (Selley 1985). No measurement of the limestone permeability has been carried out, but petrographic observation shows that the beds are highly cemented (no visible porosity). Cracking of the seal caprock and related abrupt pressure release lead to squeezing out of the mud, resulting in the fluidization of the upper part of the Olavarría Formation bearing the previously trapped acidic compaction water. Acidic properties are, of course, interpretative. The reason why they are suspected to be acidic is that the base of the limestone overlying the Olavarría Formation is undoubtedly dissolved. Calcite remains highly soluble in acidic conditions, even at low temperatures.

The high smectite content (70–90 %) of the random I/S crystallized at the expense of the ordered I/S (with <15 % smectite) of the Olavarría Formation suggests that mud injection occurred at a temperature significantly lower than the maximum temperature attained during the peak diagenesis. This fact probably implies significant uplift of the sedimentary succession before the cracking of the limestone seal caprock. The precipitation of silica (now quartz) and calcite, which occurred in the late fractures and voids of the Loma Negra limestones, attests to supersaturation of fluids with these minerals, probably in response to pressure release in the system.

## 5.1.4  Evidence of Basin Inversion

The physicochemical conditions previously mentioned would easily be satisfied during a tectonic event like a structural inversion. Basin inversion causes uplift of the sediments and erosion of a part of the sedimentary succession. Uplift and erosion are the best way to promote the cracking of the seal caprock because they increase the local pressure gradient at the boundary of the over pressured compartments (leading to the overcoming of the mechanical resistance of the seal). Basin inversion can begin at any time, depth, or temperature during the basin history and the uplift range can be highly variable according to the tectonic setting. It is not possible to accurately determine the degree of uplift when mud injection occurred in the limestone of the Loma Negra Formation.

The findings and characteristics of mud beds and pipes in the Loma Negra Formation, the direct connection of the latter with the Olavarría Formation, the transformation of ordered I/S (with <15 % smectite) into a more smectitic random I/S (90 % smectite) in the mud bed and pipes and in the unconformity between the Olavarría and the Loma Negra Formations (dissolution front) lead us to propose a paragenetic sequence for the Olavarría and the Loma Negra Formations (Table 5.4) based, in part, on Worden and Burley (2003) and Burley and MacQuaker (1992). This area was affected by a compressive tectonic stage, basin uplift, erosion, faulting, and folding of sedimentary strata. The major alteration processes consisted of fracturing and injection of over pressured fluidized sediments within the limestones of the Loma Negra Formation. These phenomena are particularly well developed near the basin margins and are reported in the literature as telogenesis (Worden and Burley 2003).

**Table 5.4** Summary of the paragenetic sequence proposed for the Olavarría and Loma Negra Formations

| Observations | | Processes | Geological history stage | |
|---|---|---|---|---|
| Minerals | Structures | | | |
| I/S (<15 % Sm) | Overlying limestone strata dissolution | Compaction fluid overpressure Organic maturation | Burial stage | |
| | Faults in the limestones | SW–NE Compression | Tectonic effect of Ventania | |
| I/S (90 % Sm) | Mud pipes with limestone blocks | Injection of over pressured mud disruption and collapse of limestone blocks | Uplift | Telogenesis |
| Secondary calcite chalcedony | Local graben (extension) | Release of fluid pressure | Erosion | |

Taken from Zalba et al. (2007)

## 5.1.5 Scanning Electron Microscopy of the Clays of the Olavarría Formation

In Fig. 5.11a, the texture of the gray clays from the base of the sequence, with a face-to-face disposition, is typical of sedimentary deposits. They are predominantly composed of detrital illite, but in addition there are diagenetic illite-smectite (I/S)

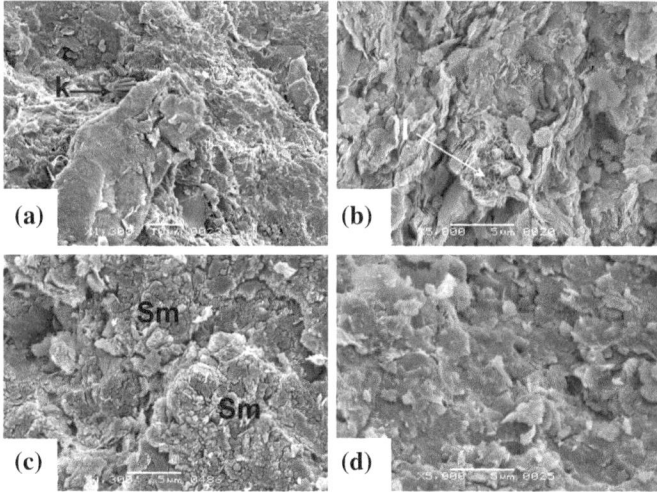

**Fig. 5.11** Scanning electron micrographs of clays of the Olavarría Formation. **a** Texture of the gray clays with Kaolinite (k) that shows a face-to-face disposition. **b** Ocher clays with the same texture that **a**, but with needle-like aggregates of diagenetic illite (I). **c** Ocher illitic clays with diagenetic smectite (Sm) showing desiccation cracks. **d** Reddish clays with face-to-face orientation of the particles and also with some development of "swirl pattern"

and kaolinite; the latter in the form of "books" and in small proportion. The ocher clays that overlie the gray sediments in the sequence offer the same type of texture, but needle-like aggregates of diagenetic illite are also observed (Fig. 5.11b). Also in the ocher clays, the presence of diagenetic smectite is observed (Fig. 5.11c), with the typical texture of this mineral and, furthermore, affected by cracks due to drying of the material which shrinks and loses water reversibly (expansive clay). The reddish clays of the upper part of the sequence, also show face-to-face disposition of the flakes and consist of illite-smectite. Figure 5.11d is an example of the texture of the reddish clays from the roof of the sequence, with face-to-face orientation of the particles and also with some development of "swirl pattern" characteristic of sedimentary deposits in marine environment. They are also composed of illite and illite-smectite.

### 5.1.6  Technological Properties of the Clays of the Olavarría Formation

Garrido et al. (1996) prepared hand-molded specimens (plastic mud) on samples of ocher and red stocks. As a first step the pieces were heated at 950, 1000, and 1050 °C. The Pyrometric Cone Equivalent (PCE) value for the red stock is 7 (1250 °C) while for the ocher stock is 9 (1280 °C). Value of the plasticity index (PI) for the red stock is associated with the content of illite and the presence of fine particles in the sample (Table 5.5).

In Table 5.6 the final contraction, water absorption and apparent porosity of the product, indicate the differences between the two materials. Red clay has high values of contraction and maximum temperature tested, the specimens are brown and they are vitrified. Vitrification temperature depends on the amount and type of mineral phases that act as fluxes. The high content of alkalis in illite decreases this temperature. The same occurs when $Fe_2O_3$ content is high. The contraction is often related to mechanisms of sintering among particles and the formation of vitreous phases during heating.

On ignition at 950 °C, it would be feasible to achieve materials for the manufacture of tiles (adsorption <15 %). When heated at higher temperatures (1000–1050 °C) specimens are little porous, but the excessive shrinkage limits its

**Table 5.5**  Plasticity index and Pyrometric Cone Equivalent (PCE) values of ocher and red "stock pile" of the Olavarría Formation

| Technological test | Red clay | Ocher clay |
|---|---|---|
| Plasticity index | 7.6 | 4.5 |
| Fraction <44μm % | 94 | 84 |
| Fraction >2 μm % | 44 | 25 |
| PCE, temperature (°C) | 1250 | 1280 |

Taken from Garrido et al. (1996)

**Table 5.6** Linear shrinkage (LS), water absorption (WA), apparent porosity (AP) and mechanical resistance (MOR) depending on the calcination temperature of ocher and red "stock pile" of the Olavarría Formation

| T (°C) | Red clay | | | Ocher clay | | |
|---|---|---|---|---|---|---|
| | LS (%) | WA (%) | AP (%) | LS (%) | WA (%) | AP (%) |
| 100 | 5.5 | – | – | 1.3 | – | – |
| 950 | 11.5 | 10 | 21.6 | 4.0 | 14.9 | 27.7 |
| 1000 | 15.4 | 4.1 | 9.6 | 7.0 | 10.7 | 21.7 |
| 1050 | 17.0 | 0.4 | 1.0 | 8.2 | 7.0 | 15.0 |
| Water content (%) | 29 | | | 23 | | |

Taken from Garrido et al. (1996)

application. It would be then appropriate to continue the study with pressed-molded specimens (low water content) to obtain products for floor and wall coverings.

The ocher clay presents less contraction and increased porosity in the heated materials with respect to the red clay products. These values are probably linked to high quartz content and coarse particle size that characterizes this variety. The percentage of carbonate recognized by chemical data, mineralogy and the existence of particle agglomerates would support these features. According to the tests, in the range of temperature between 950 and 1050 °C, ocher clay would be suitable for the preparation of red ceramic with low absorption of water (<15 %). However, it has the disadvantage of low plasticity. This was possible to solve, in part, by the addition of 3–5 % bentonite. In the latter case, (Table 5.7) the characteristics are similar to those of the natural sample, except for an increase in the contraction; although no marked decrease on the absorption of water with respect to the original material is registered.

Because of the high shrinkage presented by the red clay and in order to improve the quality of the product obtained, mixtures of both clays were prepared: ocher and red, in proportions of 13 and 20 % of red clay (Table 5.8). The results indicate that by mixing the ocher clay with 20 % of the red clay, good products can be obtained when heated at 1050 °C. Red clay helps to reduce the adsorption of water to values <5 %, with regular contraction.

In general, the clays have low plasticity index. They are associated with illite content and fine particles (clay fraction). They have low-refractoriness, and are classified as varied red clays (PCE 7–1250 °C) and varied ocher clays

**Table 5.7** Technological properties of the composite mixture by ocher clay and bentonite

| T (°C) | Ocher clay + 3 % of bentonite | | | Ocher clay + 5 % of bentonite | | |
|---|---|---|---|---|---|---|
| | LS (%) | WA (%) | AP (%) | LS (%) | WA (%) | AP (%) |
| 100 | 4.1 | – | – | 4.3 | – | – |
| 950 | 5.4 | 14.4 | 27.2 | 5.6 | 14.0 | 26.5 |
| 1000 | 6.3 | 13.0 | 25.3 | 8.3 | 11.9 | 23.4 |
| 1050 | 9.7 | 8.7 | 18.0 | 10.0 | 8.8 | 18.3 |
| Plasticity Index | 7.6 | | | 10 | | |

Taken from Garrido et al. (1996)

**Table 5.8**  Technological properties of the composite mixture by ocher and red clays

| T (°C) | Ocher clay + 13 % of red clay | | | Ocher clay + 20 % of red clay | | |
|---|---|---|---|---|---|---|
| | LS (%) | AA (%) | PA (%) | LS (%) | AA (%) | PA (%) |
| 100 | 1.2 | – | – | 1.7 | – | – |
| 950 | 3.7 | 14.5 | 27.5 | 5 | 13.8 | 26.4 |
| 1000 | 5.2 | 11.7 | 23.4 | 5.3 | 10.9 | 22.11 |
| 1050 | 9.1 | 5.5 | 12.2 | 9.6 | 3.4 | 8.0 |
| Plasticity Index | 4.7 | | | 5 | | |

Taken from Garrido et al. (1996)

(PCE 9–1280 °C). Their low plasticity requires additives to improve workability. They are used in red ceramic, both in local factories and in other important urban centers of the province and also they are used in the manufacture of Portland cement participating, along with limestone, in the preparation of green powder. There are numerous quarries in exploitation today in the Sierras Bayas area.

### 5.1.7  Sedimentary Deposits; Characteristics and Mineralogical Composition of the Cerro Negro Formation

The other geological unit of important clay deposits in the area of Sierras Bayas is the Cerro Negro Formation (see Table 1.1) which has been found in the subsoil and by the opening of several quarries near the Cerro Negro village (Fig. 5.12) located 7 km to the SSW of Sierras Bayas town and 40 km to the W of the city of Azul, in the La providencia quarry, Sierras Bayas, as well as in the town of Villa Cacique and Calera San José del Cármen, already mentioned above (Villa Cacique Sector).

**Fig. 5.12**  Cerro Negro Formation, La Providencia quarry, Sierras Bayas

**Fig. 5.13** Cerro Negro
Formation, unconformity
(*white line*) between the
Lower Depositional System
(LDS) and the Upper
Depositional System (UDS),
La Providencia quarry, Sierras
Bayas

The type area for this formation is located in the neighborhoods of the town of Cerro Negro where it was discovered for the first time by Ing. Victorio Angelelli in 1973 and characterized as the Cerro Negro Formation by Iñíguez and Zalba (1974). According to sedimentological and stratigraphic evidence, the Cerro Negro Formation was divided by Andreis et al. (1992) in two depositional systems separated by an irregular paleosuperface (Fig. 5.13). Stratigraphic sections carried out allowed these authors to infer that the Lower Depositional System (LDS, Fig. 5.14a) outcrops in the area of Sierras Bayas, near Olavarría city, while the Upper Depositional System (UDS, Fig. 5.14b) is found near the town of Cerro Negro, in different areas of Sierras Bayas, in the Calera San José del Cármen and in the region of Villa Cacique village. The LDS measuring up to 28 meters thick in the La providencia quarry, Sierras Bayas, is mainly composed of olive-gray and pink massive to laminated siltstones and claystones, and of a phosphatized black limestone breccia at the base of the unit.

The UDS measures up to 90 m in thickness in the neighborhoods of the town of Cerro Negro and comprises quartzose sandstones, siltstones, claystones, and hetherolitic facies. The shales (siltstones and claystones) with marlstone intercalations are reddish and greenish colored, with thicknesses which reaches 140 m (drilling data). They are laminated, with load casts and barite concretions (nodules) up to 10 cm long. The clays are of low plasticity, which is corrected, usually, with the addition of plastic clays, preferably bentonite. The mineralogical composition of the clays in this formation is different in the two depositional systems. Table 5.9 shows that the LDS (CSM) samples are mainly composed of illite + I/S, with abundant quartz impurities. Feldspars are not observed, but calcite is present in marls and calcareous clays.

The UDS (CNP samples) is mainly illitic, with the systematic presence of scarce smectite and interstratified chlorite-smectite (C/Sm). Impurities are composed of very abundant quartz, feldspars, and iron oxides and hydroxides (hematite and goethite).

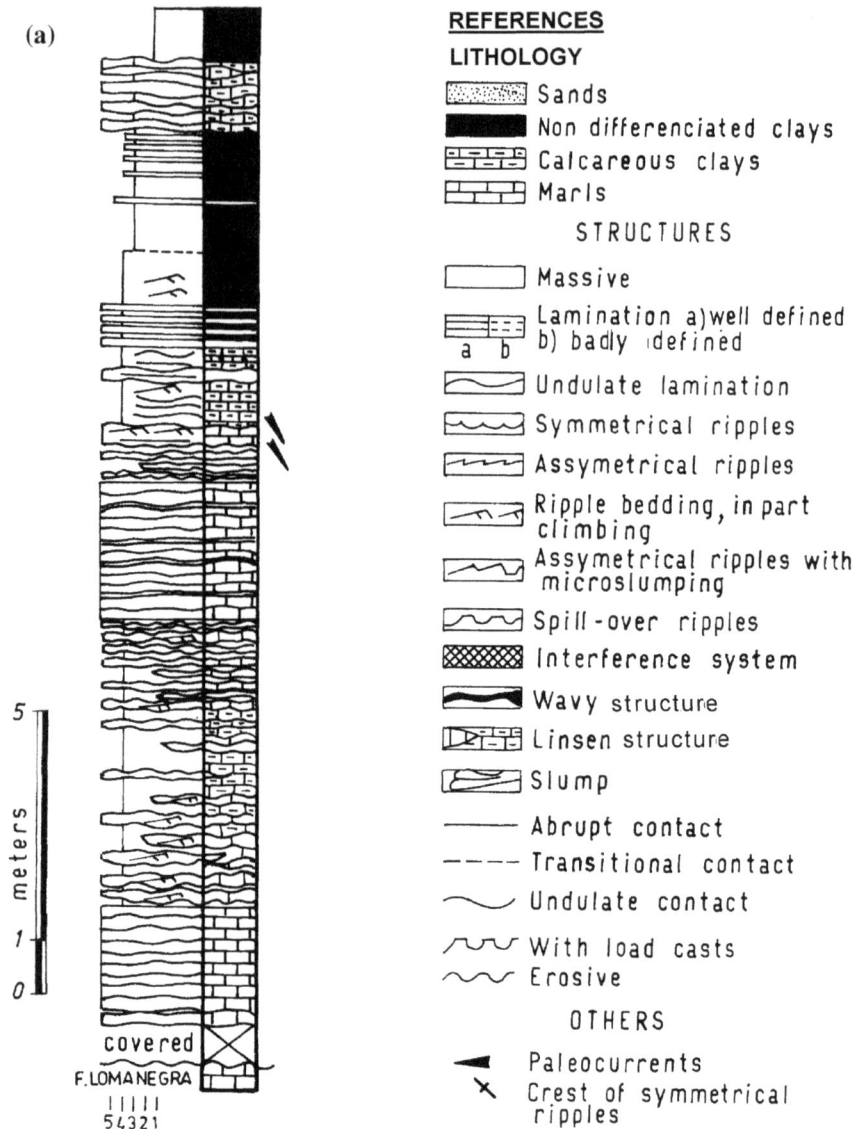

**REFERENCES**

**LITHOLOGY**

▨▨▨ Sands
■■■ Non differenciated clays
▤▤▤ Calcareous clays
▤▤▤ Marls

STRUCTURES

▭ Massive

▤▤▤ Lamination a)well defined
a   b    b) badly defined

◁▭ Undulate lamination

◁▭ Symmetrical ripples

◁▭ Assymetrical ripples

◁▭ Ripple bedding, in part climbing

◁▭ Assymetrical ripples with microslumping

◁▭ Spill-over ripples

▨▨▨ Interference system

■■■ Wavy structure

▨▤▤ Linsen structure

◁▭ Slump

──── Abrupt contact

──── Transitional contact

⌣ Undulate contact

⌒⌒ With load casts

⌒⌒ Erosive

OTHERS

◀ Paleocurrents

✕ Crest of symmetrical ripples

**Fig. 5.14**  Cerro Negro Formation, La Providencia quarry, Sierras Bayas. **a** Stratigraphic section of the Lower Depositional System (LDS) **b** Stratigraphic section of the Upper Depositional System (UDS). Taken from Andreis et al. (1992)

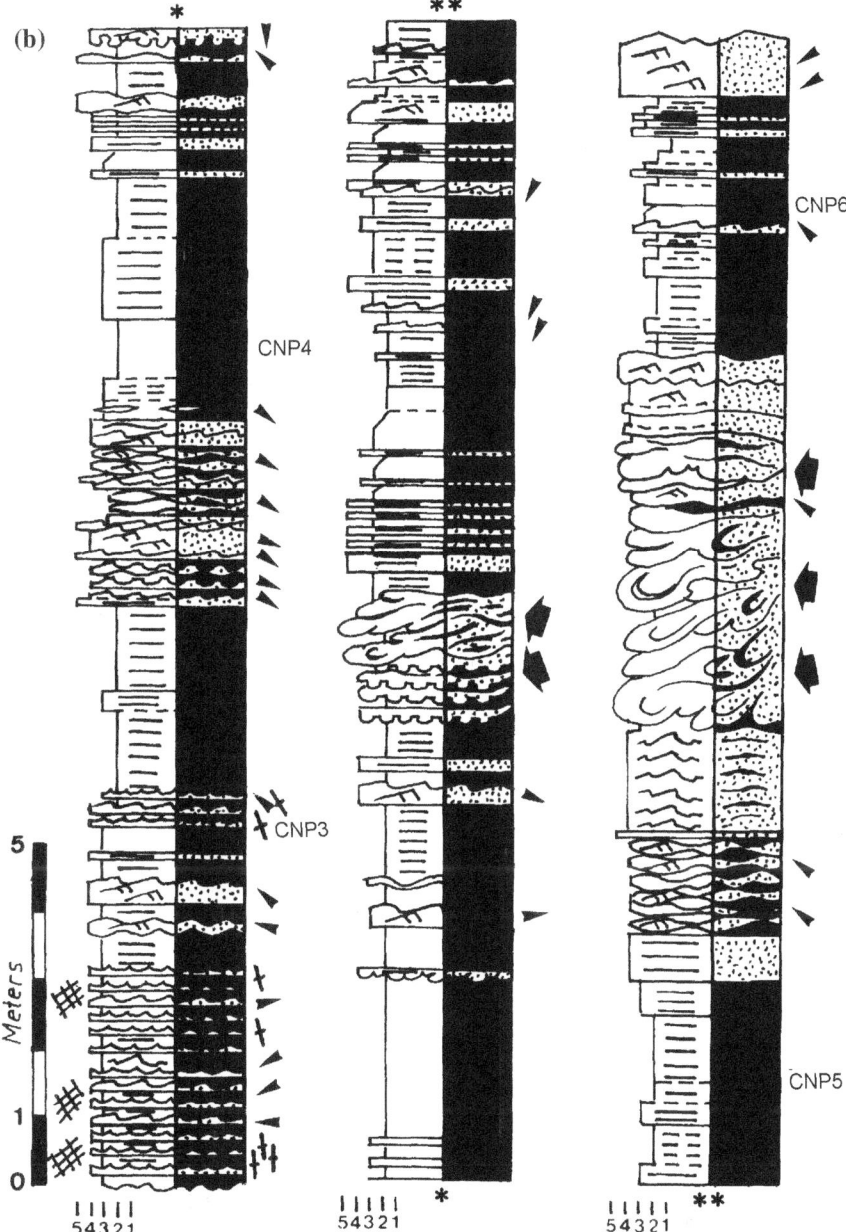

**Fig. 5.14**   (continued)

**Table 5.9** Mineralogical composition by X-ray diffraction of clays from the two depositional systems of Cerro Negro Formation

|  | Oriented samples <4 microns (%) | | | Other components (bulk sample) | | | | |
|---|---|---|---|---|---|---|---|---|
|  | Sample | IM | C/Sm | Sm | Q | F | Cal | IMT | P |
| Upper Depositional System (UDS) | CNP6 | 65 | 20 | 15 | v.a | a | – | I + IS (<15 % Sm) | Md/M |
|  | CNP5 | 65 | 25 | 10 | v.a | a | – | I + IS (<15 % Sm) | Md/M |
|  | CNP4 | 75 | 20 | 5 | v.a | a | – | I + IS (<15 % Sm) | Md/M |
|  | CNP3 | 70 | 30 | – | v.a | a | v.a | I + IS (<15 % Sm) | Md/M |
|  | CNP2 | 65 | 30 | 5 | v.a | a | – | I + IS (<15 % Sm) | Md/M |
|  | CNP1 | 70 | 25 | 5 | v.a | a | – | I + IS (<15 % Sm) | Md/M |
| Lower Depositional System (LDS) | CSM9 | 80 | 5 | 15 | v.a | – | s | I + IS (<15 % Sm) | Md/M |
|  | CSM8 | 100 | – | – | v.a | – | – | I + IS (<15 % Sm) | Md/M |
|  | CSM7 | 100 | – | – | a | – | v.a | I + IS (<15 % Sm) | Md/M |
|  | CSM6 | 100 | – | – | v.a | – | v.a | I + IS (<15 % Sm) | Md/M |
|  | CSM4 | 100 | – | – | v.a | – | – | I + IS (<15 % Sm) | Md/M |
|  | CSM3 | 100 | – | – | v.a | – | – | I + IS (<15 % Sm) | Md/M |

*References IM* illite materials; *C/Sm* chlorite-smectite; *Sm* smectite; *Q* quartz; *F* feldspars; *Cal* calcite; *IMT* illític material type; *P* polytype; *v.a* very abundant; *a* abundant; *s* scarce
Taken from Zalba et al. (1994)

## 5.1.8  Scanning Electron Microscopy of the Clays of the Cerro Negro Formation

Figure 5.15a shows a texture with a "swirl pattern," typical of sedimentary deposits in marine environment, corresponding to illitic clays of the Cerro Negro Formation (LDS). Figure 5.15b represents illitic clays of the (UDS). An aggregate of illite-smectite, with filamentous forms and diagenetic in origin grows on a basis of irregular shaped crystals.

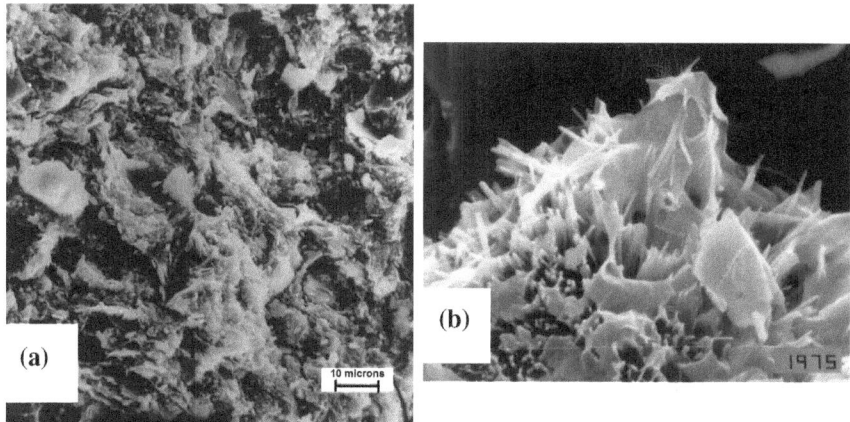

**Fig. 5.15** Scanning electron micrographs of the Cerro Negro Formation. **a** Upper Depositional System, illític clays showing a "swirl pattern" texture. **b** Lower Depositional System. Illite-smectite, with filamentous forms of diagenetic origin. *Scale* 5 μm

### 5.1.9 Technological Properties of the Clays of the Cerro Negro Formation

Technological tests were carried out in "composite" samples (stocks) for each of the depositional systems. The results can be seen in Table 5.10. LDS samples have smaller particle size and greater PI than samples from the UDS. The clays of the two depositional systems present similar equivalent cones (PCE around 15). As the main constituent is illite, associated alkali favor a low refractoriness of the clays. Pieces subjected to drying and heating tests were characterized by their linear contraction, porosity, water absorption, and flexural strength in the composite sample CNP + bentonite and in the composite sample CSM. Results for semi-dry and plastic pieces are shown in Table 5.10. After drying, the pieces show no appreciable differences. Shrinkage values were low and, therefore, indicate low green resistance. Heated samples have uniform red colors as well as good appearance, without cracks or fissures.

Linear shrinkage of the composite sample CSM is higher than that from the composite sample CNP at 1050 °C and, consequently, the water absorption is low. Similar results are obtained after heating both samples at 1100 °C. Ceramic products are characterized by a linear contraction of around 10 %, without deformation. They are dense, without porosity, although they show some signs of vitrification. These preliminary results indicate that both samples of composite clays CSM and CNP are suitable for the manufacture of red ceramic tiles with low water absorption and are classified as varied clays. They are exploited in the area of Cerro Negro locality, with excellent results and large reserves (Zalba et al. 1994) for their use in the manufacture of ceramic floor coverings.

**Table 5.10** Comparison of technological properties of "composite samples" of both depositional systems of the Cerro Negro Formation

| T °C | Upper Depositional System: CNP + Bentonite | | | | | Lower Depositional System: CSM | | | | |
|---|---|---|---|---|---|---|---|---|---|---|
| | LS (%) | AP (%) | WA (%) | AD (g/cm³) | MOR (kg/cm²) | LS (%) | AP (%) | WA (%) | AD (g/cm³) | MOR (kg/cm²) |
| Plastic mud state | | | | | | | | | | |
| 100 | 3.0 | – | – | – | 13 | 4.0 | – | – | – | 13 |
| 1000 | 5.0 | 25.5 | 13.0 | 2.0 | 147 | 4.0 | – | – | – | 13 |
| 1050 | 6.8 | 17.3 | 6.6 | 2.1 | 235 | 8.9 | 12.2 | 5.4 | 2.2 | 325 |
| 1100 | 11.4 | 2.6 | 1.1 | 2.3 | 415 | 10.9 | 5.0 | 2.1 | 2.4 | 325 |
| Semi-dry state | | | | | | | | | | |
| 100 | 0.0 | – | – | – | 33 | 0.0 | – | – | – | 20 |
| 1000 | 0.0 | 19.2 | 8.7 | 2.2 | 214 | 0.5 | 20.1 | 9.6 | 2.1 | 265 |
| 1050 | 1.0 | 12.1 | 5.2 | 2.3 | 421 | 2.9 | 12.3 | 5.3 | 2.3 | 357 |
| 1100 | 2.5 | 1.6 | 0.8 | 2.4 | 514 | 4.7 | 6.0 | 2.5 | 2.4 | 442 |

*References* $T$ temperature; $LS$ line contraction; $AP$ apparent porosity; $WA$ water absorption; $AD$ bulk density; $MOR$ modulus of rupture
Taken from Zalba et al. (1994)

# References

Andreis RR, Zalba PE (1989) Estratigrafía y paleogeografía de las secuencias cuarciticas al oeste de Barker (Buenos Aires, Argentina). In: Proceedings of 1° Jornadas Geológicas Bonaerenses, Tandil, pp 909–930

Andreis RR, Zalba PE, Iñíguez AM (1992) Paleosuperficies y sistemas depositacionales en el Proterozoico superior de Sierras Bayas, Sistema de Tandilia, Provincia de Buenos Aires, Argentina. In: Proceedings of 4° Reunión Argentina de Sedimentología, vol 1, pp 283–290

Andreis RR, Zalba PE, Iñíguez Rodriguez AM, Morosi M (1996) Estratigrafía y evolución paleoambiental de la sucesión superior de la Formación Cerro Largo, Sierras Bayas (Buenos Aires, Argentina). In: Proceedings of 6° Reunión Argentina de Sedimentología, pp 293–298

Burley SD, MacQuaker JHS (1992) Authigenic clays, diagenetic sequences and conceptual diagenetic models in contrasting basin-margin and basin-center North Sea Jurassic sandstones and mudstones. In: Houseknecht DW, Pittman ED (eds) Origin, Diagenesis and Petrophysics of Clay Minerals in Sandstones, vol 47. SEPM Special Publication, Tulsa, Oklahoma, USA, pp 81–110

Garrido L, Zalba PE, Pereira E (1996) Aplicación tecnológica de arcillas (Acopios) de la sucesión superior de la Formación Cerro Largo, Buenos Aires, Argentina. In: Proceedings of 7° Reunión Argentina de Sedimentología y 1° Simposio de Arcillas, Bahía Blanca, pp 305–310

Higgins GE, Saunders JB (1967) Report on 1964 Chatham Mud Island, Erin Bay, Trinidad, West Indies. AAPG Bull 51(1):55–64

Iñíguez MA, Zalba PE (1974) Nuevo nivel de arcilitas en la zona de Cerro Negro, partido de Olavarría, provincia de Buenos Aires. LEMIT Serie 2(264):95–100

Massabie A, Amos A, Iturriza R (1992) Diapirismo arcilítico tectoinducido, Sierras Bayas, Provincia de Buenos Aires. Rev Asoc Geol Argentina 47(4):389–398

Pickering KT, Agar SM, Ogawa Y (1988) Genesis and deformation of mud injections containing chaotic basalt limestone chert associations: examples from the southwest Japan forearc. Geology 16:881–885

Poiré DG (1987) Mineralogía y sedimentología de la Formación Sierras Bayas en el núcleo septentrional de las sierras homónimas, Olavarría provincia de Buenos Aires. Tesis Doctoral 494 (inédito). Facultad de Ciencias Naturales y Museo, Universidad Nacional de La Plata, p 271

Poiré DG (1993) Estratigrafía del Precámbrico sedimentario de Olavarría Sierras Bayas, provincia de Buenos Aires, Argentina. In: Proceedings of 13° Congreso Geológico Argentino y 3° Congreso de Exploración de Hidrocarburos, Mendoza, vol 2, pp 1–11

Sellés Martínez J (1994) Lineamientos estructurales y evolución extensional de la plataforma Neoproterozoica-Eopaleozoica de las Sierras Septentrionales de la Provincia de Buenos Aires (Argentina). Rev Brasil Geocien, Sao Paulo, 2(3):289–295

Selley RC (1985) Elements of petroleum geology. W.H. Freeman, San Francisco p 449

Von Gosen W, Buggisch W (1989) Tectonic evolution of the Sierras Australes fold and thrust belt (Buenos Aires province/Argentina). Geologisches Rundschau, Stuttgart 79(3):797–821

Worden RH, Burley SD (2003) Sandstone diagenesis: the evolution of sand to stone. In: Burley SD, Worden RH (eds) Sandstone diagenesis: recent and ancient. Reprint Series of the International Association of Sedimentologists, Blackwell Publishing Ltd., New York, pp 3–44

Zalba PE, Iñiguez AM, Volzone C, Morosi M (1996) Mineralogía y procesos postdeposicionales en la sucesión superior de la Formación Cerro Largo, Sierras Bayas, (Buenos Aires, Argentina). In: Proceedings of 7° Reunión Argentina de Sedimentología y 1° Simposio de Arcillas, Bahía Blanca, pp 299–304

Zalba PE, Volzone C, Garrido L, Morosi M, Pereira E (1994) Mineralogical composition and diagenetic processes in the two depositional systems of the Cerro Negro Formation, Buenos Aires, Argentina: industrial application. Rev Geol Chilena 21(2):303–311

Zalba PE, Manassero M, Laverret EM, Beaufort D, Meunier A, Morosi M, Segovia L (2007) Middle Permian telodiagenetic processes in Neoproterozoic sequences, Tandilia System, Argentina. J Sediment Res 77:525–538

Zalba PE, Morosi ME, Manassero M, Conconi MS (2010) Microscale diagnostic diagenetic features in Neoproterozoic and Ordovician units, Tandilia basin, Argentina: a review. J Appl Sci 10(22):2754–2772

# Chapter 6
# General Pueyrredón, Balcarce, and Necochea Counties

**Abstract** In the Sierra Bachicha and Sierra del Volcán, near the Balcarce city, clay deposits are the result of basement rock hydrothermal alteration. The two best-known ones are called Cerro Segundo and María Eugenia, respectively. At María Eugenia mine the basement rocks have been obliterated by kaolinite formation and it is only possible to recognize the original foliation in reduced sectors. Cerro Segundo deposits also revealed the dominant presence of kaolinite, in this case, as a product of superimposed weathering argillization. Clays from Cerro Segundo are hard, with low plasticity, high content of alumina, and white cooking color. They are classified as refractory clays. They have been regularly used as an additive to lime. The Balcarce Formation outcrops discontinuously in almost all the Tandilia System, except in the Sierras Bayas Sector, but it is better exposed in the SE extreme of the range. Based on trace fossils its age is considered Cambrian-Early Ordovician. It may overlay uncomformably, weathered basement rocks; the Metapelitas Punta Mogotes or either sediments of the Las Aguilas or the Cerro Negro formations. The clay strata intercalated in the Balcarce Formation reaches only 1 m in thickness in the eastern area of Tandilia and are recovered, in general, as a by-product of the exploitation of quartzite levels, historically used for typical constructions of the zone (and named "Mar del Plata stone"). The mineralogical composition of the Balcarce Formation clays is fundamentally kaolinitic. The San Ramón clay deposits are classified as Flint Clays. They are intercalated between thick quartzite strata (Balcarce Formation). The mineralogical composition is mainly kaolinitic, with dickite associated in minor proportions. These clays are hard, whitish, plastic, or refractory (San Ramón), the latter used as chamotte and for the production of refractory materials. Based on previous research a geodynamic evolution of the Tandilia basin is offered.

**Keywords** Mar del Plata (Chapadmalal)-Balcarce-Necochea Sector · Geology · Stratigraphy · Basement · Hydrothermal · Residual · Sedimentary deposits · Cambrian-Ordovician trace fossils · Mineralogy · Kaolinite · Dickite · Technology · Geodynamic evolution

## 6.1  Mar del Plata (Chapadmalal)-Balcarce-Necochea Sector

### 6.1.1  Residual Deposits; Characteristics, Mineralogical and Chemical Composition

#### 6.1.1.1  Cerro Segundo and María Eugenia

In the Sierra Bachicha and Sierra del Volcán, in the neighborhoods of the Balcarce city, clay deposits are the result of basement rock hydrothermal alteration. The two best-known ones are called Cerro Segundo and María Eugenia, respectively (see Fig. 1.1). The Cerro Segundo site is located on the eastern slope of the Sierra de Bachicha (Balcarce), km 61, route 226, which connects the cities of Mar del Plata and Balcarce. Residual white kaolinitic clays have originated from gneiss rock types.

Geological studies on the area started with Schiller (1938) and subsequently continued with Di Paola and García Espiasse (1986), and Angelelli and Garrido (1988). These authors preliminarily attributed the origin of the kaolin to hydrothermal solutions that altered the original feldspars from the circulation of thermal fluids through what they interpreted as a horizontal fault. Yet, the surface is more likely to represent an unconformity (a paleosurface, a great hiatus) that separates the basement rocks from the overlying quartzites of the Balcarce Formation.

María Eugenia mine, exploited in the seventies and at present inactive, is located 20 km east of the city of Balcarce, and 400 m far from the route 226 (km 47.5), on the western slopes of the Sierra del Volcán. This kaolin mine was an open pit operated through two tasks. According to Delgado et al. (2011) old exploitation of kaolin in the María Eugenia mine has an area of approximately 150 m long by 60 m wide. The basement rock presents varying degrees of argillization. In the East sector of the mine the rock keeps an E–W foliation, crisscrossed by quartz veins. In the western part of the quarry the structure of the basement has been obliterated by kaolinite formation and it is only possible to recognize the original foliation in reduced sectors. Iron and oxy-hydroxides veins are linked with the development of quartz and kaolin.

X-ray diffraction analyses on Cerro Segundo deposits carried out by Di Paola and García Espiasse (1986) revealed in the clay fraction the dominant presence of kaolinite with illite + I/S; associated smectite in varying proportions, and impurities of quartz. $Al_2O_3$ content varies between 21 and 38 % and those of $Fe_2O_3$ between 1 and 3 % (Iñiguez and Zalba 1988).

At María Eugenia, up to 70 % of kaolinite, with illite + I/S and smectite in minor proportions (5–15 %) and impurities of quartz were recorded. The tenors of $Al_2O_3$ vary between 22 and 37 % and those of $Fe_2O_3$ between 2.6 and 4 % (Iñiguez and Zalba 1988).

Delgado et al. (2011) focused in the geological and mineralogical knowledge of the kaolinitic clays from the Sierra del Volcán, which were in the past exploited at the María Eugenia mine. The authors applied modern techniques of study such as Reflectance Spectroscopy SD Field Spec ProTM, SEM and the use of X-Powder software for the interpretation of X-ray diffraction patterns, which allowed them to recognize different zones within the mineral deposit with varying degrees of argillization. According to them, to the west of the deposit they identified the association kaolinite—of high crystallinity—and illite $2M_1$ along with the development of a dense texture of kaolinite crystals, which would indicate a hydrothermal origin. Also, they detected subordinate kaolinite of low crystallinity and less dense textures associated with goethite which would indicate a supergene origin. The presence of sulfides and quartz veinlets cross-cutting the Complejo Buenos Aires rocks, recognized in the western area, would be, according to the authors, evidence of fluid flowing through this basement rocks and could be the responsible for the developed argillic alteration.

### 6.1.2  Scanning Electron Microscopy of Residual Deposits of Cerro Segundo

Figure 6.1 is important because it shows "books" of incipient kaolinite growing from feldspar, very rare phenomenon to document during the development of diagenetic or weathering processes. In this case, this process is the beginning of the argillic alteration by weathering effects and this type of kaolinite correspond to the weathering process superimposed to the original hydrothermal ones occurred in the residual deposits of Sierra de Bachicha.

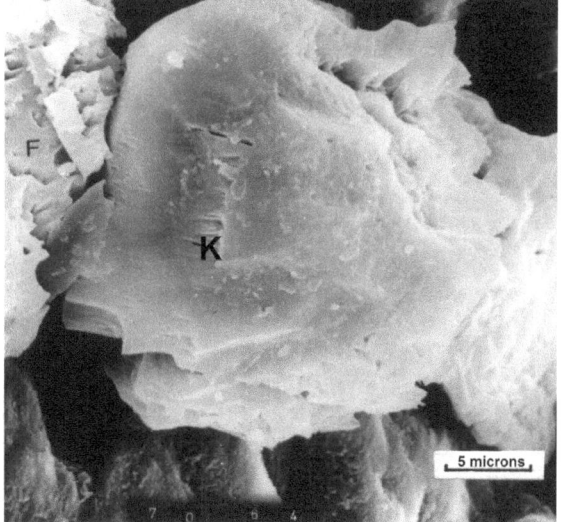

**Fig. 6.1** Cerro Segundo. "Books" of incipient kaolinite (K) growing from feldspars (F). Taken from Zalba et al. (1988)

### 6.1.3   Technological Properties of Residual Deposits of Cerro Segundo

Clays from Cerro Segundo are hard, with low plasticity, high content of alumina and white cooking color. Because of the tenors of $Al_2O_3$ and $Fe_2O_3$ and their technological behavior they are classified as refractory clays. They have been regularly used as an additive to lime.

### 6.1.4   Sedimentary Deposits; Characteristics and Mineralogical Composition of the Balcarce Formation

From Blanca Grande locality to Mar del Plata area (along 230 km) a series of sediments consisting of sandstones, sabulites, orthoconglomerates of siliceous composition, together with intercalated siltstones and claystones, correspond to the Balcarce Formation. The Sierra del Volcán Diamictite, previously assigned to the Neoproterozoic (Iñiguez 1999), is considered of Ordovician age, based on detrital zircon age dating (Van Staden et al. 2010) and also it is considered as a member of the Balcarce Formation (see Table 1.1). Poiré et al. (2003) assigned the Balcarce Formation to the Cambrian-Early Ordovician on the basis of trace fossils. Rapela et al. (2007) reported a maximum depositional age of 475–480 Ma based on SHRIMP analyses over detrital zircon grains. The sandstones dominate, while the sabulites, conglomerates, siltstones, and claystones are subordinate (Fig. 6.2). According to Poiré and Spalletti (2000) the geometry of the sandstone beds is

**Fig. 6.2** View of the Balcarce Formation quartzites, Sierra de La China. (Courtesy Marcelo Manassero)

sheet-like; most sedimentary bodies are bounded by convex-upward surfaces, though some wide channel-like features are also present. Planar and tangential cross-stratifications are the dominant structures within sandstone bedsets, and large-scale sigmoidal bodies are frequent in most sections. Sheet-like and lenticular sandstone–mudstone interbeds are commonly intercalated among sandstone storeys. Trace fossils are abundant at the top surface of the sandstone member in sandstone–mudstone interbeds. The quarries all around Batán and Chapadmalal towns allow depicting the stratigraphic architecture of the Balcarce Formation. Tidal processes are inferred from the features of cross-bedded sandstone facies (bars) and heterolithic (wavy and lenticular) facies (swales). Large to medium scale laterally persistent bodies of cross-bedded sandstones exhibit rhythmic lateral variations in the thickness of foresets and in clay content due to spring and neap tide alternation. Clay drapes covering foresets and other sedimentation surfaces herringbone cross-bedding, opposite palaeocurrent trends in successive sedimentary bodies and reactivation surfaces also suggest tidal deposition. The migration and accretion of bidimensional sand bars seem to be controlled by highly asymmetrical time-velocity tidal currents. Subordinated, high-energy storm episodes are suggested by hummocky cross-bedded sandstones, sheet conglomerates armouring previous tidal sand bodies and heavy mineral concentrations in the wavy sandstone laminae of heterolithic facies.

These sediments outcrop, discontinuously, in almost all the Tandilia System, except in the Sierras Bayas Sector, but they are better exposed in the SE extreme of the sierras: Balcarce and Mar del Plata areas, with thicknesses reaching up to 90 m (the largest of all the outcropping mountain range formations). The Balcarce Formation culminates in the SW sector at Cabo Corrientes, Mar del Plata city, which represents the extension of the sierras into the Atlantic Ocean (Fig. 6.3). The outcrops of the Balcarce Formation at Sierra de Los Padres can be seen in Fig. 6.4.

In the northwestern zone (El Ferrugo-Constante 10-El Cañón Sector, and Chillar Sector) and in the central area of the sierras (Villa Cacique Sector; La Juanita, Sierra de La Tinta, Cuchilla de Las Aguilas Sector) the Balcarce Formation shows its base

**Fig. 6.3** Last south-eastern outcrops of the Balcarce Formation, Cabo Corrientes, Mar del Plata

**Fig. 6.4** Quartzites of the Balcarce Formation, Sierra de los Padres, Mar del Plata

exposed and with few meters in thickness (between 10 and 15 m). The Balcarce Formation may overlay, uncomformably, weathered basement rocks—Complejo Buenos Aires—(e.g. El Ferrugo, Constante 10, and Chillar); the Metapelitas Punta Mogotes (in the area of Mar del Plata) or either overlay sediments of the Las Aguilas Formation (in the La Juanita-Cuchilla Las Aguilas Sector) or deposits of the Cerro Negro Formation (in the Villa Cacique Sector).

The Balcarce Formation shows abundant trace fossils and a high ichnodiversity. Zalba et al. (1988) revised the ichofossils described by different authors in the whole Tandilia System pointing their own findings like vertical burrows (Endichnia) crossing planar cross-bedded sandstone facies in tidal environments (Fig. 6.5a); Dydymaulichnus sp. Young, in rippled sandstone facies (Fig. 6.5b); Plagiogmus sp., Rœdel; Scolicia sp., De Quatrefages. Crimes (verbal communication 1986) thought that the description made by Borrello (1966) of *Crossopodia scotica* in Los Pinos quarry corresponded to *Cruziana furcifera*, D′Orbigny. *Cruziana* is considered a valuable chronostratigraphic trace fossil. It is the most important fossil described in the Ordovician strata of the North of Argentina

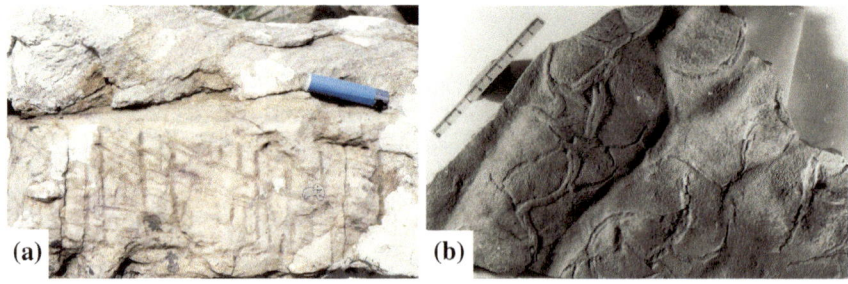

**Fig. 6.5 a** Vertical burrows (Endichnia) crossing planar cross-bedded quartzites (tidal), Balcarce Sector, Balcarce Formation. **b** Ichnofossils (Didymaulichnus sp.) in quartzites with ripple marks of the Balcarce Formation. Photograph **b** taken from Zalba et al. (1988)

(Aceñolaza 1978) and considered as indicative of a lower Ordovician age in all Europe and Bolivia. Particularly, the association of *Arthrophycus*, *Cruziana* and *Skolithos* is typical of the Portugal lower Ordovician and *Skolithos* and *Cruziana* are distinctive fossils of the Lower Ordovician of Jujuy and Salta, Argentina, according to Borrello (1966). Also *Didymaulichnus* is characteristic of the Ordovician of Europe, though it already is present from the Upper Cambrian (for more details cf. Zalba et al. 1988).

After revising the already published material thoroughly and taking into account new discoveries made by Poiré and Spalletti (2000) and Poiré et al. (2003) these authors considered that an updated list of trace fossils of the Balcarce Formation found by different authors is: *Ancorichnus, Arthrophycus allegha niensis, Arthrophycus isp., Bergaueria isp., Cochlichnus isp., Conostichuis isp., Cruziana furcifera, Cruziana isp., Daedalus labeckei, Didymaulichnus lyelli, Didymaulichnus isp., Diplichnites isp., Diplocraterion isp., Herradurichnur scagliai,? Monocreterion isp., Monomorphichnus isp., Palaeophycus alternatus, Palaeophycus tubularis, Palaeophycus isp., Phycodes aff. pedum, Phycodes isp., Plagiogmus isp., Planolites isp., Rusophycus isp., Scolicia isp.* and *Teichichnus isp.*

Poiré et al. (2003) found U-shaped, perpendicular to bedding burrows containing spreiten between the vertical tubes; the burrows are usually seen at their intersections with the upper surface of the sandstone beds. They appear as dumbbel-shaped burrows, consisting of paired circular openings of the vertical burrows, joined by a slit-shaped area of disturbed sediments corresponding to the spreiten (References: *Diplocraterion isp.*, Borrello (1966), and Poiré et al. (2003) for Parque San Martín, Mar del Plata, and Dazeo quarries (Don Mariano and La Gloria), Batán locality, 15 km from Balcarce city.

The clay strata intercalated in the Balcarce Formation reaches only 1 m in thickness in the eastern area of Tandilia. They are whitish to greenish colored, and are recovered, in general, as a by-product of the exploitation of quartzite levels, historically used for typical constructions of the zone (and named "Mar del Plata stone"). Figure 6.6 shows a front of a quarry in Batán locality, where minor clay levels are intercalated in the quartzites and which were in the past subjected to intense exploitation. Conversely, at San Ramón quarry (see Fig. 1.1) clay levels have been the main object of exploitation.

The mineralogical composition of the Balcarce Formation clays is fundamentally kaolinitic, with lower proportions of illite, illite-smectite (I/S) and scarce impurities of quartz. X-ray diffraction analysis, performed on total samples at: La Barrosa, Los Pinos (Balcarce), La Florida, Iacusa and Defeudo (Chapadmalal) quarries can be found in Table 6.1. In some cases, the composition is 100 % kaolinite, and scarce smectite (diagenetic), and pyrophyllite (detritic) traces have also been found.

San Ramón I and II quarries are important Fire-Clay deposits situated in the Cerro El Tigre, 18 km to the NE of the town of Claráz (Necochea). Exploitation works are located in two well-defined areas separated by a ground road that leads to the La Numancia village. The southern sector has been exploited by combining open pit and underground mining, using cameras and pillars, while the north (exhausted), was only exploited underground. The surface of this mining property

**Fig. 6.6** Batán quartzite quarry, where intercalated clays in the Balcarce Formation were mined

**Table 6.1** Mineralogical composition of clays by X-ray diffraction from different quarries of the Balcarce Formation: La Barrosa and Los Pinos (Balcarce); La Florida, Iacusa and Defeudo (Chapadmalal)

| Quarry<br>Mineral | La Barrosa | Los Pinos | La Florida | Iacusa | Defeudo |
|---|---|---|---|---|---|
| Kaolinite | 65 | 66 | 70 | 72 | 90 |
| Illite+I/S | 25 | 24 | 20 | 22 | 5 |
| Smectite | – | – | 3 | 4 | 2 |
| Quartz | 10 | 10 | 7 | 2 | 3 |

Taken from Iñíguez and Zalba (1974)

reaches 294 hectares. The San Ramón clay deposits form a compact mudstone bank, pale gray colored, with conchoidal fracture (Fig. 6.7a) and with thicknesses varying between 1.50 and 2.50 m. They are intercalated between thick quartzite strata (Balcarce Formation). From the study of thin sections on the clays of San Ramón carried out for this contribution it can be seen that the rock is a mudstone, predominantly composed of kaolinite, in some zones recrystallized to dickite (bigger grain size and of higher birefringence than kaolinite) and with "greenish eyes", as well as minor illite of higher birefringence than kaolinite. The rock is laminated with grains oriented with their long axis parallel to former bedding planes (Fig. 6.7b, c). In the upper right part of the photograph (Fig. 6.7b) various grains are in contact showing that they have experimented compaction. Illite is disposed as tangential clay coatings around grains (Fig. 6.7c), in this case around big hematite crystals as can be seen in a detail of the former photograph (Fig. 6.7d), showing that it is detrital. Monocrystalline and polycrystalline quartz grains are well rounded, with undulate extinction (lower right corner), typical of crystalline basement rocks and showing their provenance. A spindle-shaped net of microbial mats (in black) once carbonaceous and now hematized are observed with their long axe parallel to earlier bedding planes (Fig. 6.7e). These algal mats are very similar to those

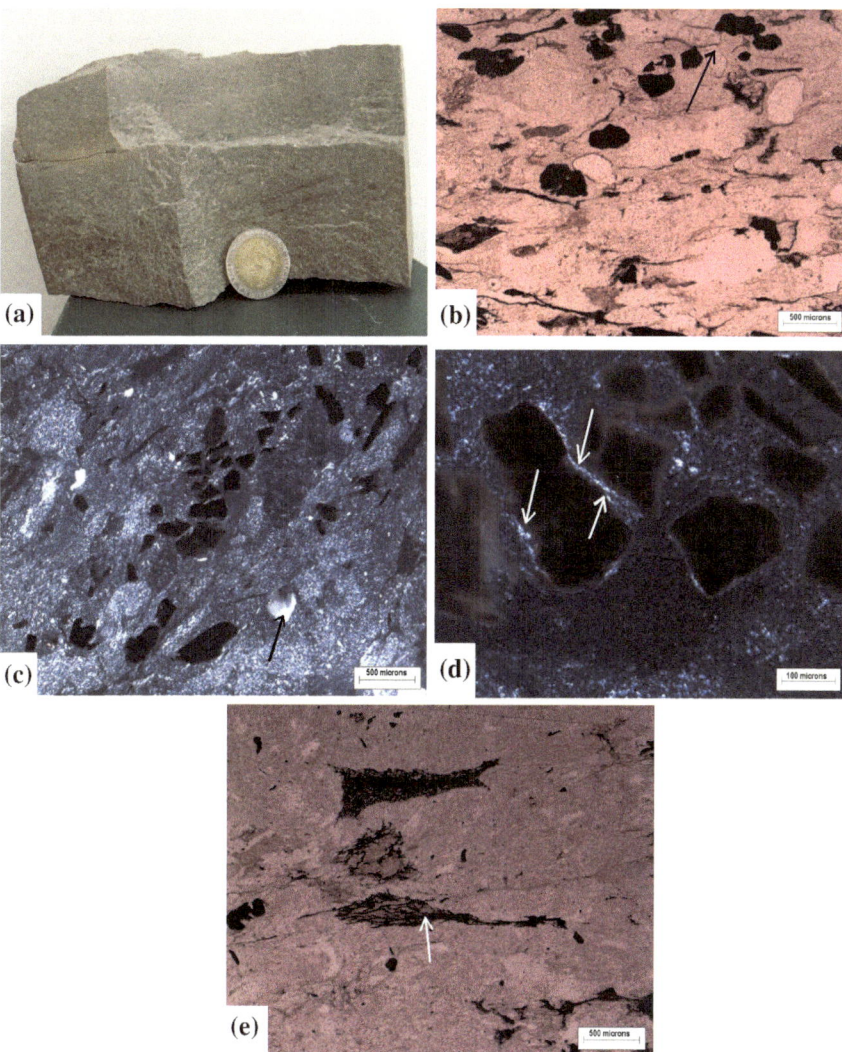

**Fig. 6.7  a** San Ramón Flint-Clay. Dense, hard, greenish kaolinite-dickite sedimentary clay with choncoidal fracture. Balcarce Formation. **b** Laminated claystone. Grains oriented with their long axis parallel to former bedding planes. In the upper right part various grains are in contact showing that they have experimented compaction (*arrow*) PPL. **c** Detrital illite tangential clay coatings around hematite crystals. Well rounded, monocrystalline quartz with undulate extinction (*arrow*) XPL. **d** Detail of illite tangential clay coatings (*arrows*) XPL. **e** A spindle-shaped net of microbial mats (*arrow*) PPL

described in Middle Proterozoic siliciclastic deposits by Schieber (2007) (see Fig. 5c, d, Chap. 5: Newland Formation of the Belt Supergroup, Montana, and Rampur Shale of Vindhyan Supergroup, India, respectively). Consequently, the San Ramón clays are classified as siliciclastic deposits with MISS.

**Table 6.2** Technological tests of San Ramón clays (PCE) Pyrometric Cone Equivalent

| Sample | PCE | Temperature PCE (°C) | Linear shrinkage (°C) | | Liquid limit | Plastic limit | Plasticity index |
|--------|-----|----------------------|-----------------------|-------|--------------|---------------|------------------|
| San Ramón | 35 | 1785 | 1100 | 3.5 % | 13 % | 15 % | 2 % |
| | | | 1500 | 6.3 % | | | |

The Balcarce Formation clays are hard, whitish, plastic and refractory, used as chamotte and for the production of refractory materials (Domínguez and Schalamuk 1999; Lopez et al. 2002). Selected technological analyses on clays of San Ramón, Balcarce, carried out for this work, can be seen in Table 6.2.

The physico-chemical and technological properties of the clays of San Ramón point out to refractory characteristics and within these, to the so named "Flint Clays", because they resemble a flint. They are very fine-grained clays, primarily composed of kaolinite, ordered or well ordered, which breaks with a characteristically conchoidal fracture and have a bulk density of 2.2–2.5. Typically, "flint clays" have experimented "digestion" process by which the original texture of the rock has been obliterated. Clay seems to have been digested by the formation of a gel from which would crystallize a new kaolinite. When the reaction leads to the formation of a gel the resulting clay is a "flint". These clays are characterized by a micro to cryptocrystalline texture of clay minerals that are arranged like a tight mosaic of interwoven grains (Keller 1968, 1976). Nevertheless, based on Dristas and Frisicale (1987, 1996), Zimmermann and Spalletti (2005) interpreted these deposits as reworked pyroclastic sediments either transported primary by wind and then reworked in the Balcarce basin or related to nearer volcanic centers. However, the authors do not locate these nearer volcanic centers. Recently, Cingolani et al. (2010) presented U–Pb zircon ages by ID-TIMS on samples taken from kaolinitized pyroclastic levels (*sic*) outcropped in Cerro del Corral (according to Dristas and Frisicale 1987). The studied zircon crystals show mainly idiomorphic bipyramidal type characteristics, suggesting acidic igneous origin and reveal a very few transport from the source. The analyzed zircons indicate much older ages than the Ordovician-Silurian age of Balcarce sedimentation. Most of $207Pb/206Pb$ ages are close to 2.1 Ga, suggesting a main Paleoproterozoic source area for these crystals, representing the typical age of the Tandilia basement. Among the 18 analyzed zircon fractions, extracted from volcaniclastic layers (Cingolani et al. 2010), no evidence of an Ordovician-Silurian volcanic episode was obtained, suggesting that these crystals represent a reworking product of previous Paleoproterozoic volcanic rocks.

Based on isotopic data of kaolinite, Domínguez and Silleta (2002) attributed the genesis of these deposits to weathering processes.

Carin 1, Carin 2, and Mody mines: this group of mines is located 9 km from the La Negra town, on the south western slopes of the Sierra del Piojo, Necochea, and 1.7 km from the local road that leads to Barker town. Exploitation was performed by open pit mining, in two areas, Carin 1 and Carin 2. The clay strata are from 3 to 6 meters thick, covered by approximately 12 m thick of quartzite deposits. The area occupied by these mining properties reaches 72 hectares. According to Domínguez and Ullman (2005) "at Loma del Piojo the argillization reaches 26 m thick and extends 900 m along a hillside and decreases with depth". The clay is soft, with red, yellow, and white hues, irregularly distributed and it presents remnant textures of volcanic type (porphyric). Clays are covered by the Balcarce Formation and its stratigraphic position is uncertain because the deeper debris extracted ($-26$ m) corresponds to a diabase whose stratigraphic position is unknown (basement or the Balcarce Formation?; Rapela et al. 1974). A description of the geology can be found at Domínguez and Silleta (2002). The mineralogy has a zoned distribution. At surface the content of kaolinite is greater, whereas smectite and hematite increase with depth. An average mineralogical composition is 14 % illite, 74 % kaolinite, 12 % quart and traces of smectite. At the site three types of clay have been classified which could be commercially used as a kaolinitic type: with kaolinite >70 % and smectite <20; an intermediate type: with kaolinite between 30 and 60 % and smectite <60 %; and a bentonitic type: with smectite type >70 % and kaolinite <20 %, according to the latest mentioned authors.

## 6.1.5  Scanning Electron Microscopy of the Balcarce Formation Clays

In Fig. 6.8a, the texture of clays from La Barrosa (Balcarce) quarry presents an apparent bimodal distribution where hexagonal platy kaolinite predominates. In the same quarry, Fig. 6.8b shows the texture of kaolinitic clay, with face-to-face disposition of the plates and some "swirl" texture. Plates of larger size are of illite. Figure 6.8c corresponds to clays from the Los Pinos quarry (Balcarce) where illite shows large sheets. The clays of Chapadmalal show a texture with face-to-face orientation plates. Kaolinitic aggregates of crystals show inbound edges caused by diagenetic dissolution (Fig. 6.8d).

The quartzites of the Balcarce Formation also bear clays in the pore rocks. An example of this is Fig. 6.8e, in the area of Balcarce city, where pores are completely filled by hexagonal crystals of kaolinite forming "books", and to a much lesser extent, aggregates of illite-smectite, with the typical texture named "wrinkly handkerchief". Both textures are unmistakable characters of diagenetic origin. San Ramón clays show a "swirl pattern", typical texture of sedimentary deposits (Fig. 6.8f). The sample is mainly composed of kaolinite with minor proportions of dickite. In Fig. 6.8g, a very locked and dense texture with "books" of diagenetic kaolinite, with a sinusoidal curve disposition can be observed.

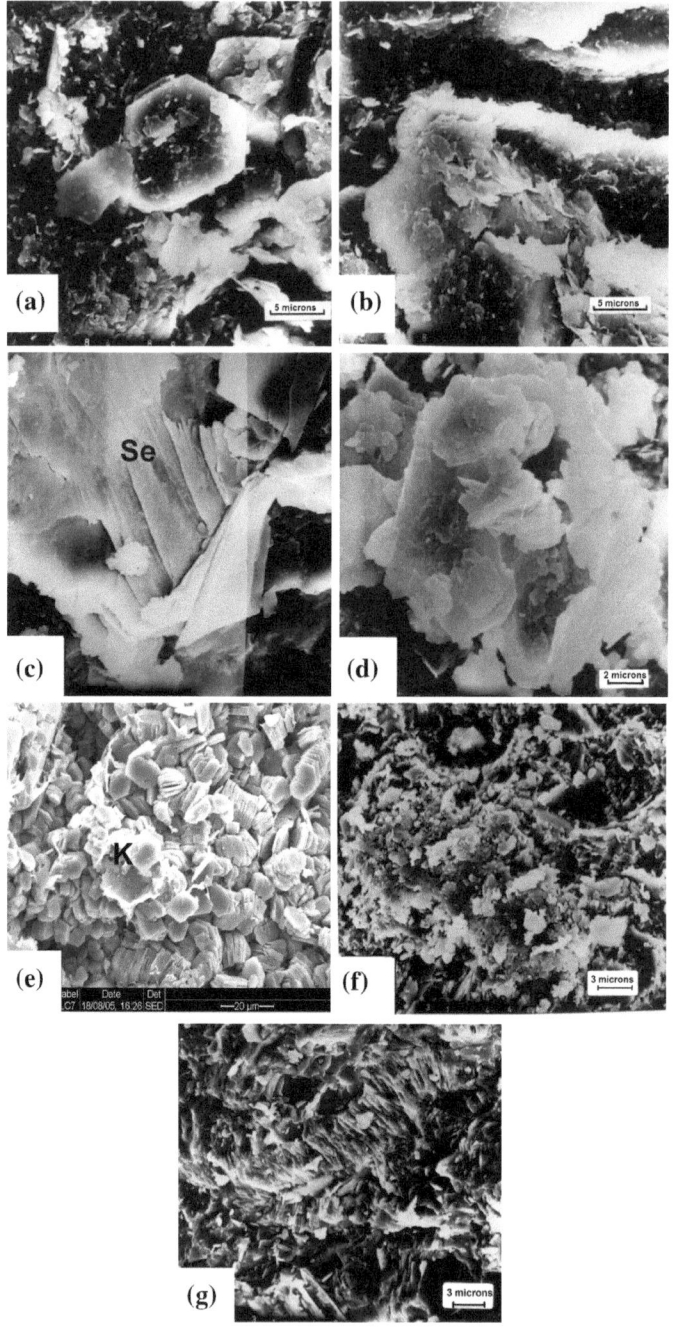

◄ **Fig. 6.8** Scanning electron micrographs of the Balcarce Formation. **a** Bimodal distribution where hexagonal platy kaolinite predominates, La Barrosa quarry. **b** Texture of a kaolinitic clay, with face–to–face disposition of the plates and some with "swirl" pattern. **c** Sericite sheets (Se), Los Pinos quarry (idem scale photo **d**). **d** Aggregates of kaolinite crystals show inbound edges (diagenetic dissolution) with face–to–face disposition. **e** Pore full filled with hexagonal crystals of kaolinite forming "books", La Barrosa quarry. **f** Clays show a "swirl pattern" texture. **g** Very locked and dense texture with "books" of diagenetic kaolinite, San Ramón quarry

## 6.1.6 Technological Properties of Sedimentary Clays of the Balcarce Formation

The plasticity of these clays is very low; the material is very sandy, with no consistency, being virtually not plastic. At 1100 °C, there is no sintering of the specimen; conversely, at 1500 °C a solid cylinder could be obtained. The specimens were prepared with water and with Arabic gum in order to be able to mold them. Regarding the clays of the Loma del Piojo, according to Domínguez and Ullman (2005) kaolinitic clays have a plasticity index of 17; of 21 for the intermediate clays, and of 23 for the bentonitic clays. Because of their grain size and ceramic properties they are suitable for coatings with acceptable properties of molding. Due to their alumina content they have good fire resistance. In a certain zone of the quarry, red clay is exploited in open pits. This material is relatively refractory, plastic and with little quartz content. Clays of this site have not been incorporated in common ceramic bodies, being its main problem the variation of their mineralogy and color. In some zones of the site, a good geological control would allow clear clays to be extracted. The hill presents excellent exploration potential.

## 6.1.7 Geodynamic Evolution of the Tandilia Basin

The results offered by Zalba et al. (2007) provide an additional stage in the tectonosedimentary evolution proposed by Iñiguez et al. (1989). In his model (Fig. 6.9a), Iñiguez et al. (1989) proposed that from 900 to 700 Ma the sedimentary sequences experimented only weak epeirogenic movements. Consequently, the sediments were not deformed and remained flat-lying. The orogenic movements produced by the Brazilian cycle (600 Ma) affected the crystalline basement and the sedimentary cover as well producing faulting with important vertical and horizontal displacement in response to SW oriented stress. A series of horst and graben structures were formed. In some cases, the upthrown blocks were completely eroded. The crystalline basement rocks were exposed while the depressed ones preserved most of the sedimentary cover, as occurred in the areas of Sierras Bayas, Barker, and San Manuel (60 km to the SE of Barker). Presumably, the peneplanation process occurred during the Cambrian. At the same time, the basement of the

(a)

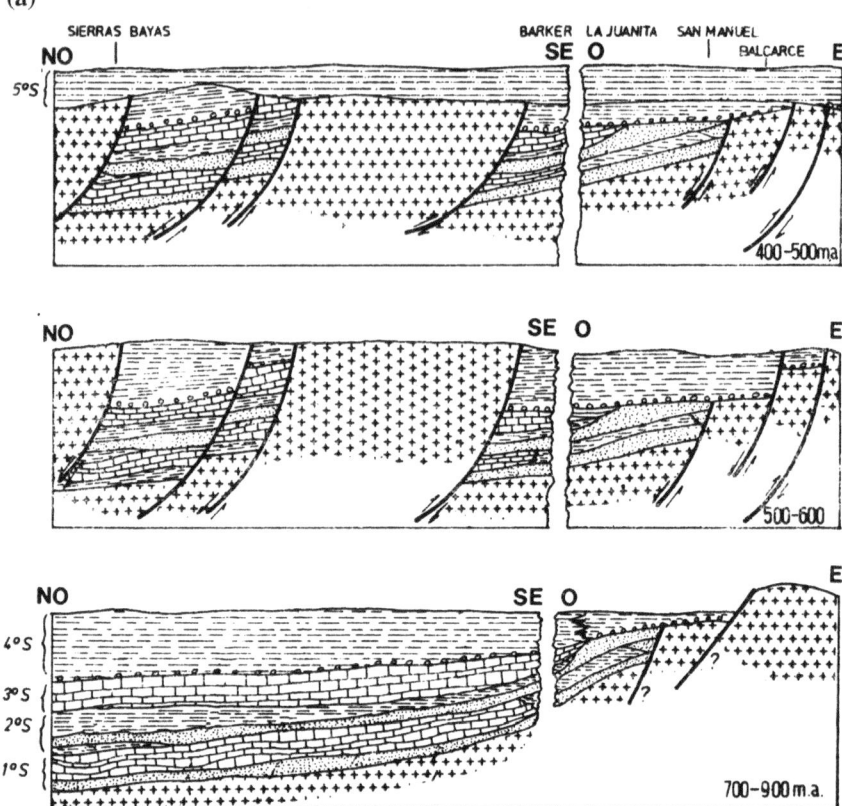

**Fig. 6.9** **a** Scheme of the tectonosedimentary evolution of the Tandilia Basin from Sierras Bayas to Balcarce areas. Taken from Iñiguez et al. (1989). 1°S: First sedimentary sequence: Cuarcitas Inferiores and dolostones. 2°S: Second sedimentary sequence: Cuarcitas Superiores and overlying claystone and siltstone facies: Cerro Largo Formation. 3°S: Third sedimentary sequence: Loma Negra Formation. 4°S: Fourth sedimentary sequence: Las Aguilas–Cerro Negro formations. 5°S: Fifth sedimentary sequence: Balcarce Formation. **b** New proposal for the tectonosedimentary evolution of the Sierras Bayas and Barker–Cuchilla de Las Aguilas areas,Tandilia Basin, representing a telodiagenetic stage in the model of Iñiguez et al. (1989), which occurred during the middle Permian (around 254 Ma). The sketch is composed of two sections: A–B (Sierras Bayas–Barker) and B–C*: (Barker–Cuchilla de Las Aguilas). For a better understanding and a clear graphic representation, section B–C has been revolved 135° (C*) on a plane following the A–B section (see references on Fig. 6.9b)

Balcarce-Mar del Plata area (100 and 150 km to the SE of Barker, respectively) experienced subsidence, having been a positive area, at least, during the deposition of the Neoproterozoic units. Shallow burial (2–3 km) took place between approximately 400 and 250 Ma.

An additional stage in the tectonosedimentary evolution of the Sierras Bayas and Barker-Cuchilla de Las Aguilas areas is presented in Fig. 6.9b, based on Zalba et al.

**(b)**

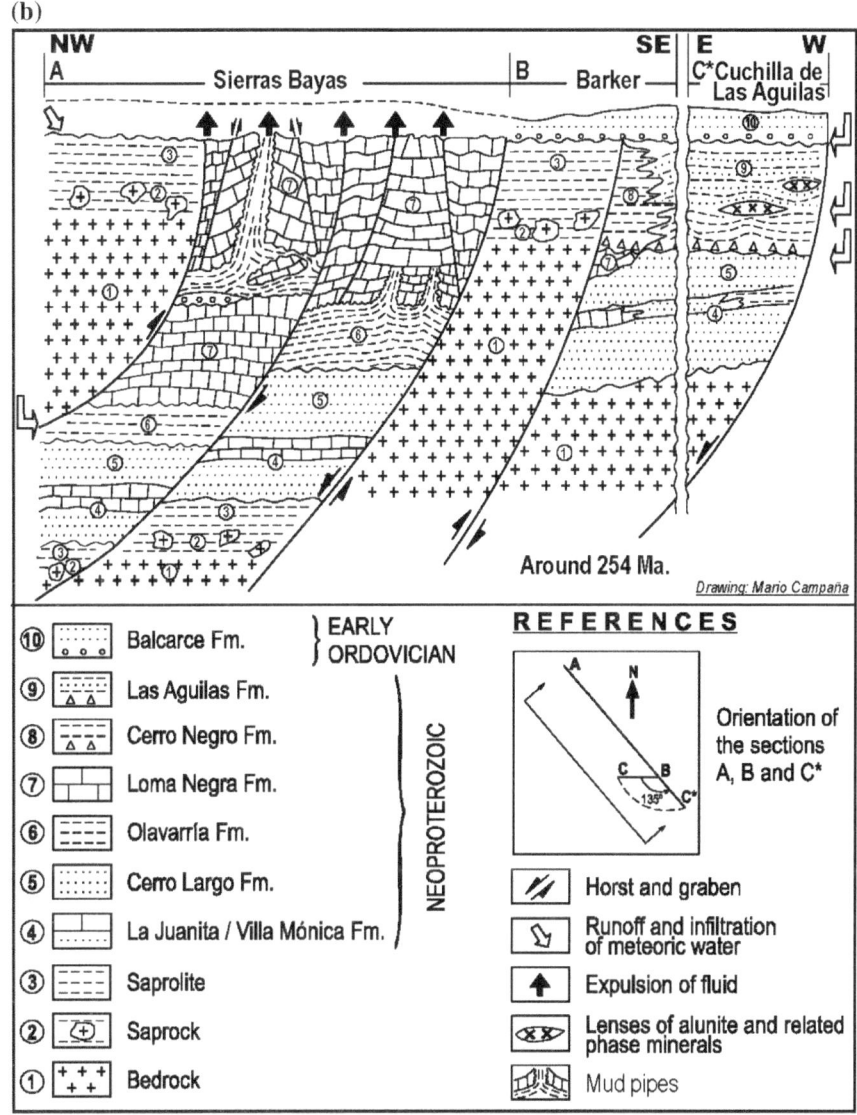

**Fig. 6.9** (continued)

(2007). A telogenetic stage occurred during the Middle Permian (near 254 Ma) in response to a renewed basin inversion, uplift, and erosion of the upper sedimentary units of the basin. Consequent transfer (and possible mixing) of fluids of both diagenetic and meteoric origin were involved along structural discontinuities previously created by the compressive regional forces and reopened by decompression. Based on alunite dating, it should be noted that the new age obtained by Zalba et al.

(2007) for the alteration processes agrees well with the Permian age proposed by Von Gosen and Buggisch (1989) and Varela et al. (1985) for the main deformation and folding stage of the Ventania System.

Vertical movements have prevailed since the Precambrian (Brazilian orogenic cycle) and during most of the geological history of this blockmountain system. Because of this, the sedimentary cover has probably been reburied and reexhumed several times.

Based on drilling data, Lesta and Sylwan (2005) assumed that Tandilia was a positive area subsequent to the Ordovician. On the basis of this information and the age of alunite in the Las Aguilas Formation, Zalba et al. (2007) also postulate that after the uplift and deformational phase occurred during the middle Permian, the Tandilia System has remained as a positive area.

# References

Aceñolaza FG (1978) El Paleozoico inferior de Argentina según sus trazas fósiles. Ameghiniana 15:15–64

Angelelli V, Garrido L (1988) Arcillas y caolines bonaerenses. Su composición y ensayos físicos. In: Proc 3° Congreso Nacional de Geología Económica, vol 3, pp 59–88

Borrello AV (1966) Trazas, restos tubiformes y cuerpos fósiles problemáticos de la Formación La Tinta, Sierras Septentrionales de la Provincia de Buenos Aires. Comisión de Investigaciones Científicas de la Provincia de Buenos Aires. Paleontografía Bonaerense 5:1–42

Cingolani CA, Basei MAS, Uriz NJ (2010) U-Pb (ID-TIMS) zircon ages on pyroclastic events from Balcarce Formation, Tandilia System, Argentina: unreworked or reworked origin? In: Proc 7° South American symposium on isotope geology, CD Rom version, Brasilia

Delgado ML, Rodríguez ME, Tessone M, Etcheverry R (2011) Estudio Mineralógico del depósito de caolín María Eugenia, Tandilia, Argentina. Ediciones Universidad de Salamanca Stud Geol Salmant 47(2):195–205

Di Paola E, Garcia Espianesse A (1986) Génesis del yacimiento de caolín Cerro Segundo, Sierra de Bachicha, partido de Balcarce, Provincia de Buenos Aires. Argentina. Rev Asoc Arg-Min Petrol Sedim 17(1–4):47–54

Domínguez E, Schalamuk I (1999) Recursos minerales de las Sierras Septentrionales, Buenos Aires. In: Zappettini E (ed) Recursos Minerales de la República Argentina, SEGEMAR, Buenos Aires. Anales, vol 35, pp 183–190

Domínguez E, Silleta A (2002) Los elementos traza, las tierras raras e isótopos estables en la determinación de la génesis de caolinita: El caso del yacimiento Loma del Piojo, prov. de Buenos Aires. Argentina. In: de Brodtkorb MK, Koukharsky M, Leal P (eds.) Mineralogía y Metalogenia, Buenos Aires, pp 127–134

Domínguez E, Ullman R (2005) Arcillas e industria cerámica. In: De Barrio R, Etcheverry R, Caballé M, Llambías E (eds) Proc 16° Congreso Geológico Argentino, La Plata. Asociación Geológica Argentina, Buenos Aires, pp 397–408

Dristas JA, Frisicale MC (1987) Rocas piroclásticas en el sector sudoeste de las Sierras Septentrionales de la Provincia de Buenos Aires. Rev. Asoc Min Petrol Sedim 18(1–4):33–45

Dristas JA, Frisicale MC (1996) Geochemistry of an altered pyroclastic suite interbedded in the sedimentary cover of the Tandilia area, Buenos Aires Province, Argentina. Zbl. Geol. Paleont., Stuttgart, Teil I, H. 7/8, pp 659–675

Iñiguez AM (1999) Cratón del Río de la Plata, 2: La cobertura sedimentaria de Tandilia. Anales del Instituto de Geología y Recursos Minerales (SEGEMAR) 29:101–106

Iñiguez AM, Zalba PE (1974) Geología de arcillas refractarias de la Provincia de Buenos Aires. República Argentina. Rev Asoc Geol Argentina 29(3):304–310

Iñiguez MA, Zalba PE (1988) Arcillas, su distribución, características y aplicaciones en Argentina. Revista Panorama Minero, Buenos Aires 125:4–17

Iñiguez AM, Del Valle A, Poiré D, Spalletti L, Zalba P (1989) Cuenca Precámbrica-Paleozoico inferior de Tandilia, Provincia de Buenos Aires. In: Chebli G, Spalletti LA (eds) Cuencas sedimentarias argentinas. Instituto Superior de Correlación Geológica, Universidad Nacional de Tucumán, Serie Correlación Geológica, vol 6, pp 245–263

Keller WD (1968) Flint clays and flint facies. Clays Clay Miner 16:113–128

Keller WD (1976) Scan electron micrographs of kaolins collected from diverse environments of origin. IV. Georgia kaolins and kaolinzing source rocks. Clays Clay Miner 24:107–113

Lesta P, Sylwan C (2005) Cuenca de Claromecó. In: Chebli G Cortin˜as J, Spalletti L, Legarreta L, Vallejo E (eds) Fronter Exploratoria de la Argentina. In: Proc 6° Congreso de Exploración desarrollo de Hidrocarburos, 10, Mar del Plata, pp 217–231

Lopez K, Botto IL, Etcheverry R (2002) Geología y mineralogía de las arcilitas localizadas en las Estancias La Rosalía, San Eduardo y Sierra de los Barrientos, Provincia de Buenos Aires. In: de Brodtkorb MK, Koukharsky M, Leal P (eds) Proc 6° Congreso de Mineralogía y Metalogenia. Buenos Aires, pp 239–246

Poiré DG, Spalletti LA (2000) Secuencias silicoclásticas y carbonáticas del Precambrico y Paleozoico inferior del Sistema de Tandilia, Argentina. Proc 2° Congreso Latinoamericano de Sedimentologia y 8° Reunión Argentina de Sedimentología. Guía de Campo, Mar del Plata, pp 1–39

Poiré DG, Spalletti LA, del Valle A (2003) The Cambrian-Ordovician siliciclastic platform of the Balcarce Formation (Tandilia System, Argentina): Facies, trace fossils, palaeoenvironments and sequence stratigraphy. Geologica Acta 1:41–60

Rapela C, Dalla Salda LH, Cingolani C (1974) Un filón básico ordovícico en la Formación La Tinta, Sierra de los Barrientos Provincia de Buenos Aires. Rev Asoc Geol Argentina 29(3): 319–331

Rapela CW, Pankhurst RJ, Casquet C, Fanning CM, Baldo EG, González-Casado JM, Galindo C, Dahlquist J (2007) The Río de la Plata Craton and the assembly of SW Gondwana. Earth Sci Rev 83:49–82

Schieber J (2007) Microbial Mat son Muddy substates—Examples of Possible Sedimentary Features and Underlying Processes. In: Schieber J, Bose PK, Eriksson PG, Banerje Sarkar S, Wladyslav Altermann W, Catuneanu O (eds) Atlas of Microbial Mats preserved within the Siliciclastic Rock Record, 117–134 pp

Schiller W (1938) Hallazgo de caolin en una falla de la Sierra Vulcan, cerca de Balcarce (provincia de Buenos Aires). Notas del Museo de La Plata Geología 3:6

Van Staden A, Zimmermann U, Gutzmer J, Chemale F Jr, Germs G (2010) Correlation of Ordovician diamictites from Argentina and South Africa using detrital zircon dating. J Geol Soc London 167:217–220

Varela R, Dalla Salda L, Cingolani C (1985) La edad Rb-Sr del Granito de Vela, Tandil. In: Proc 1° Jornadas Geológicas Bonaerenses (Tandil) Comisión Investigaciones Científicas, provincia de Buenos Aires, La Plata, Argentina, pp 881–891

Von Gosen W, Buggisch W (1989) Tectonic evolution of the Sierras Australes fold and thrust belt (Buenos Aires province/Argentina). Geologisches Rundschau 79(3):797–821, Stuttgart

Zalba PE, Andreis RR, Iñiguez AM (1988) Formación Las Aguilas, Sierras Septentrionales de Buenos Aires, nueva propuesta estratigráfica. Rev Asoc Geol Argentina, Buenos Aires 43(2): 198–209

Zalba PE, Manassero M, Laverret EM, Beaufort D, Meunier A, Morosi M, Segovia L (2007) Middle Permian telodiagenetic processes in Neoproterozoic sequences, Tandilia System, Argentina. J Sediment Res 77:525–538

Zimmermann U, Spalletti L (2005) The provenance of the lower Palaeozoic Balcarce Formation (Tandilia System, Buenos Aires Province, Argentina). In: Proc 16° Congreso Geológico Argentino, La Plata, vol 3, pp 203–210

# Glossary

**Alcali-semi-flint clay** Semi-flint clay is intermediate between flint and plastic clay in physical characteristics and clay mineral composition. The semi-flint clay consists of mixtures of kaolinite, illite, and mixed-layer clay minerals. Alkali clays refer to clays with low melting point, due to the presence of alkali, or less commonly, to the presence of alkaline earth elements

**Algal mats (Stromatolites)** Layered organo-sedimentary structures formed by the activity of cyanobacteria (blue-green algae) that develop in shallow marine subtidal to supratidal environments. The algae cover the sediment surface and themselves trap sediment to produce a laminated alternation of dark, organic-rich algal layers and organic-poor sediment layers

**Alunite (Alumstone). $KAl_3(SO_4)_2(OH)_6$** As a secondary mineral it is found as a result of altered feldspar-rich rocks

**Anatase** It crystallizes in the tetragonal system. Low-temperature polymorph of $TiO_2$. It is found as a minor constituent in igneous and metamorphic rocks and veins and druze in pegmatites. It is also a common mineral in detrital sedimentary rocks

**Angular unconformity** A surface that separates younger strata from older eroded and inclined strata, representing a gap in the geological record

**Authigenic (Often used to contrast with allogenic)** The term refers to minerals of sedimentary rocks generated or formed in situ during or after deposition

**Ball clays** An extremely fine-grained, plastic, sedimentary kaolinitic clay. Although ball clay contains considerable organic matter, it fires white or near white in color. It is usually added to porcelain and whiteware bodies to increase plasticity

**Banding** Common type structure in different types of rocks composed of tight and parallel bands with different composition, color, texture, etc

© Springer International Publishing Switzerland 2016
P.E. Zalba et al., *Gondwana Industrial Clays*,
Springer Earth System Sciences, DOI 10.1007/978-3-319-39457-2

**Basement** The term is informally used to differentiate dominantly igneous and metamorphic rocks of the Precambrian or Paleozoic in age from the layered sedimentary rocks above. The basement beneath a sedimentary basin provides the foundation onto which sediments are deposited

**Bentonite** A clay rock derived from alteration of volcanic ash. It is composed mainly of smectite (expansive clay) and has the ability to increase several times its volume when contacted by water

**Bioturbation** Removilization of sedimentary deposits by living organisms

**Breccia** Clastic sedimentary rocks composed by angular fragments (larger than 2 mm in diameter) joined by fine-grained matrix and/or mineral cement

**Cement** Substance formed by chemical precipitation from pore fluids that act as a binder between clastic components of a sedimentary rock, generally of carbonatic or silica composition

**Chlorite** A group name of Magnesium Iron Aluminum Silicate Hydroxide minerals. A silicate (phyllosilicate) mineral group associated with low-medium temperature metamorphic rocks or hydrothermal deposits. An iron-aluminum-magnesium-silicate hydroxide mineral

**Clast** Fragment of rock or mineral of any shape or composition originated from disintegration caused by physical weathering of pre-existing rocks. It is the mechanical transport unit

**Clastic dike** Structure formed by clastic material injection, which derived from underlying or overlaying rocks. The formation of a clastic dike can be considered a natural hydraulic fracturing example, where injecting of an overpressurized fluid with grains immersed in it produces the formation of a fracture in the rock invaded. Thus, the fracture propagates while fluid pressure is maintained, and when it decreases, intrusion ends

**Clay** A natural material mainly composed of fine-grained minerals, which is generally plastic at appropriate water contents and will harden when dried or fired. As a size term, the clay fraction is smaller than either 2 or 4 microns

**Clay minerals** Refers to phyllosilicate minerals which impart plasticity to clay and which harden upon drying or firing. Clay minerals are hydrous aluminum phyllosilicates, which form in the surface (soils and sediments) or subsurface (diagenesis or hydrothermal alteration products). Clays are fine-grained (less than 2 μm in size). Therefore, they require special analytical techniques for identification: X-ray diffraction, and scanning electron microscopy. Their structures are arranged in sheets which are bonded together to form layers. Structurally, the clay minerals are composed of planes of cations, arranged in sheets, which may be tetrahedrally or octahedrally coordinated (with oxygen), which in turn are arranged into layers often described as 2:1 if they involve units composed of two tetrahedral and one octahedral sheet or 1:1 if they involve units of alternating tetrahedral and octahedral sheets. Additionally some 2:1 clay minerals have interlayers sites between successive 2:1 units which may be occupied by interlayer cations, which are often hydrated. The planar structure of clay minerals give rise to characteristic platy habit of many and to perfect cleavage, as seen for example in larger hand specimens of micas. The classification of the phyllosilicate clay minerals is based collectively, on the features of layer type (1:1 or 2:1), the dioctahedral or trioctahedral character of the octahedral sheets (i.e. 2 out of 3 or 3 out of 3 sites occupied), the magnitude of any net negative layer charge due to atomic substitutions, and the nature of the interlayer material

**Claystone** Clastic sedimentary rock mainly composed of clay-sized particles. It has a massive structure, without fissility, therefore does not include the term shale

**Conformity** Relationship between sedimentary strata that have their parallel contacts in relation to the structure of the oldest rocks

**Conglomerate** Clastic sedimentary rock formed of rounded clasts larger than 2 mm in diameter

**Contraction** It is the reduction of the dimensions of a refractory material during manufacture. Firing shrinkage relates generally to linear changes in the dimensions of the refractory material

**Craton** An ancient and stable area of continental crust composed of Precambrian crystalline rock

**Cristobalite** High-temperature polymorph of silica ($SiO_2$)

**Cryptocrystalline** Very fine mineral aggregate whose individual crystals are too small to be distinguished in a petrographic microscope

**Cutan** The term is proposed by Brewer (1960) for a broad group of pedological features including coating material that covers the surface of grains and voids within soil. Cutans are composed of soil colloids that become deposited in soils by transfer mechanisms (illuviation, diffusion). When composed of clays they are named argillans. A cutan comprising an intimate mixture of clay and iron oxy-hydroxides is called ferriargillan

**Diagenesis** The physical, chemical and biological reactions occurred within sediments after it has been deposited and during burial. Diagenetic changes (dissolution, recrystallization, authigenesis, etc.) contribute to the transformation of sediments in sedimentary rocks

**Diaspore. AlO(OH)** A main constituent of bauxite and an important source of aluminum. Also it is present in high-alumina refractory clays

**Dike** A discordant tabular body of igneous intrusion that cuts across the structure (bedding or foliation) of the host rock

**Disconformity** An erosional surface that separates horizontal strata below from horizontal strata above

**Dolomite. CaMg(CO$_3$)$_2$** A common rhombohedral carbonate mineral which is the main constituent of dolostones

**Dolomitization** Process by which limestone is converted to dolostone by replacement of calcium by magnesium, usually through contact with magnesium-rich water

**Dolostone** A carbonate sedimentary rock which contains more than 50 % of mineral dolomite (CaMgCO$_3$)

**Endogenous** Processes and products resulting from the thermal energy from inside the earth

**Erosion** The process by which sediments are removed by transporting agents, causing wear between the particles transported and on the surface which are mobilized

**Eustatics** Downward or upward movements of the sea level

**Exogenous** It applies to the processes occurring on the surface of the earth

**Fabric** Orientation of the elements of a rock in space. In sedimentary rocks the fabric is primary, while in metamorphic rocks it is determined by deformation. When there is a definite orientation, the fabric is anisotropic and when the elements are randomly oriented it is isotropic

**Facies (Sedimentary)** The overall characteristics of a rock unit (color, bedding, composition, texture, fossils, and sedimentary structures) that reflect its depositional origin and differentiate the unit from others around it

**Fault** Surface of rock rupture along which the blocks on either side have moved relative to one another parallel to the fracture surface

**Feldspars** A group of common rock-forming minerals composed of aluminum and silica (aluminosilicates) with potassium, sodium, and calcium. There are two groups of feldspar: alkali feldspar (which ranges from potassium-rich to sodium-rich) and plagioclase feldspar (which ranges from sodium-rich to calcium- rich)

**Flint clay** Described as a fine-grained, compact, non-fissile, essentially monomineralic kaolinitic rock that breaks with a conchoidal ("flinty") fracture (Keller 1991). It has almost no natural plasticity, and resists slaking. Its name comes from the similarity of the fracture of the flint (or chert). Pure or relatively pure flint clays should have certain refractory properties: high PCE (Pyrometric Cone Equivalent), good bond and high structural strength Keller (1968)

**Fire clay** A kaolinitic-rich clay with excellent refractory properties

**Formation** A mappable rock unit with relatively homogeneous lithological characteristics which allows it to be distinguished from adjacent formations

**Glauconite** An iron potassium phyllosilicate mineral (mica group), normally found in dark green rounded pellets of a sand grain size. It is considered a diagnostic element indicative of continental shelf marine depositional environments with slow rates of accumulation

**Gneiss** A high-grade metamorphic rock that represents the last stage in the metamorphism of rocks before melting. It is characterized by the alternation of light and dark bands differing in mineral composition. The lighter bands contain mostly quartz and feldspar, the darker ones often contain biotite, hornblende, garnet or graphite

**Goethite. FeO(OH)** It is an oxyhydroxide of iron by product of weathering of iron-bearing minerals like magnetite, pyrite and siderite

**Graben (See horst)** A block of crust which moves down between two parallel normal faults

**Granite** A coarse-grained felsic, intrusive igneous rock, composed of quartz and alkali-feldspar minerals and small amounts of ferromagnesian minerals (biotite or hornblende)

**Granitoid** The general term used for all intrusive igneous rock composed of quartz and feldspar

**Granodiorite** An intrusive rock similar to granite in composition, but plagioclase feldspar is more abundant than orthoclase feldspar

**Group** Lithostratigraphic unit ranks that are higher than the formation. They are composed of two or more related formations

**Halloysite. $Al_2SiO_5(OH)_4 \cdot 2H_2O$** A 1:1 aluminosilicate clay mineral, similar in structure to kaolinite, although the halloysite usually shows tubular crystal shapes. The halloysite loses interlayer water very easily, and so, shows partial dehydration

**Hematite. $Fe_2O_3$** It is a common rock-forming mineral found in sedimentary, metamorphic, and igneous rocks. Hematite is the principal ore of iron

**Hornblende** A black or dark green mineral that forms needle-shaped or fibrous crystals it is the most common member of the amphibole group and occurs in both igneous and metamorphic rocks

**Horst (See graben)** A block of crust which moves up between two parallel normal faults

**Hydrotermal** Pertaining to hot water, to the action of hot water, or to the products of this action, such as a mineral deposit precipitated from a hot aqueous solution, with or without demonstrable association with igneous processes; also, said of the solution itself. Hydrothermal is generally used for any hot water. Sometimes it is restricted to water of magmatic origin

**Igneous** Related to rocks or minerals formed by the cooling and crystallizing of molten material (magma)

**Illite** Term that refers to clay-sized micaceous mineral with a basal spacing of 10 amstrongs. It is a nonexpanding, dioctahedral, 2:1 phyllosilicate that posses a smaller layer charge and potassium (K) content than muscovite. Illite is common clay mineral in sedimentary rocks, soils and sediments. It also forms by alteration of micas, feldspars and other clay minerals

**Illite-Smectite mixed-layer** Mineral formed by stacking layers of illite and smecite

**Intraclast** Clastic particle of partially lithified sediment, derived from erosion of intrabasinal sediments

**Intracratonic basin** Basin formed within the boundaries of a craton and bounded by normal faults

**Joint** A fracture in a rock along which there has been no displacement

**Kaolin** Claystone composed mainly of a mineral of the Kaolin group (kaolinite, nacrite, dickite, or halloysite). The term also refers to a sub-group of phyllosilicates that are dioctahedral, with 1:1 layers and the layer charge is near zero. It is used as a raw material in the manufacture of ceramics, refractories and paper

**Kaolinite.** $Al_2Si_2O_5(OH)_4$ Very common clay mineral (Kaolinite-serpentine group) produced by weathering or hydrothermal alteration of feldspars and other aluminosilicates

**Karst** Type of relief due to the dissolution of limestone by rainwater laden with carbon dioxide

**Limestone** Sedimentary rock composed by more than 50 % of calcium carbonate, originated by inorganic or organic processes

**Lithofacies** A mappable subdivision of a stratigraphic unit distinguished by its observable sedimentary features (composition, texture, structure, etc)

**Magma** Molten rock material generated in the interior of the earth by melting material temperature above 600 °C. The cooling and crystallization give rise to magmatic or igneous rocks

**Marl** A sedimentary rock containing a mix of clay and limestone. The rocks can be both clastic and chemical-biogenic origin

**Matrix** Fine-grained material surrounding larger grains in a sedimentary rock

**Member** A member is the formal lithostratigraphic unit next in rank below a formation and is always a part of some formation

**Montmorillonite** A member of the Smectite Group

**Metamorphism** The process of mineralogical and structural changes of pre-existing rocks in solid state due to high temperature and/or high pressure

**Metasomatism** Process that involves introduction of materials leading to replacement of minerals

**Mica** A general term for platy phyllosilicates of 2:1 layer without swelling capability. This group includes Muscovite and biotite among other mineral species

**Micrite** Microcrystalline calcite, with grain size finer than 4 microns. It may be of detrital origin, inorganic or organic (biogenic) precipitation

**Microsparite (See sparite)** Sparry calcite in 5–20 microns size range

**Migmatite** A heterogeneous rock composed of igneous and metamorphic materials formed by partial melting (anatexis). It contains light bands called leucosome (feldspar, muscovite and quartz) and darker bands called melanosome (biotite and amphibole)

**Mixed-layer minerals** Minerals formed by stacking, regular or not, of different layers of clay minerals (Holtzapffel 1985)

**Mudrock** Fine-grained (less than 0.064 mm) siliciclastic sedimentary rock that constitutes 50 % of all sedimentary rocks of the geological record. The different types of mudrocks include: mudstone, siltstone, claystone and shale

**Mullite. $3Al_2O_3 \cdot 2SiO_2$** A solid solution phase of alumina and silica commonly used in refractory ceramic. It has creep resistance and high compressive strength at high temperature

**Nodule** A rounded or irregular mineral concretion that has a different composition from the surrounding sedimentary rock that encloses it. The nodules may be of siliceous composition, carbonatic, phosphate, iron, manganese or clay among the most common

**Ordovician** Period of the Paleozoic Era from about 485–443 million years ago

**Outcrop** A deposit or rock body exposed at the earth's surface. In many cases the bedrock may be covered by soil and vegetation, making it difficult to observe. However, in areas where the surface covering has been removed by erosion or tectonic uplift the rock may be exposed

**Oxidation** A process that involves the chemical reaction of a substance with oxygen

**Paleosol** A horizon of ancient soil, preserved by burial beneath layers of rock or more recent sediments

**Paleozoic** Era of geologic time from about 540–245 million years ago. It includes the Cambrian, Ordovician, Silurian, Devonian, Carboniferous and Permian periods

**Paragenesis** A characteristic association of minerals with a common origin. The term also refers to the order of formation of associated minerals in time succession

**Pellet** Spherical to ovoid aggregates composed of carbonate mud (micrite), lower than 0.1–0.3 mm in diameter, without internal structure

**Petrography** This technique allows the description and systematic classification of rocks by examination of thin sections

**Phyllosilicates** A wide family of minerals contain continuous two dimensional tetrahedral sheets of composition $T_2O_5$ (T: Si, Al) with tetrahedral linked by sharing three corners of each, and with a fourth corner pointing in any direction. The tetrahedral sheets are linked in the unit structure to octahedral sheets, or to groups of coordinated cations, or individual cations

**Plagioclase. $CaAl_2Si_2O_8–NaAlSi_3O_8$** Group of Feldspars which is one of the most important rock-forming minerals. They show a complete compositional range between calcium-rich plagioclase (anorthite) and sodium-rich plagioclase (albite)

**Plastic fire clay** Refractory clay which is plastic when wet

**Plasticity index** Numerical difference between the liquid and plastic limits. It represents the range of moisture contents in which the material behaves plastically

**Plutonic** Igneous rocks (also called intrusive) solidified deep within the Earth's crust, which have coarse-grained textures

**Polymictic conglomerate** Conglomerate composed of clasts from different origin

**PPL** Plane polarized light

**Precambrian** Geological period extended from the origin of the earth (about 4.5 billion years ago) to the beginning of the Paleozoic era (about 540 million years ago)

**Pseudomorphism** It occurs when a mineral is altered in such a way that its internal structure and chemical composition is changed but its external form is preserved

**Pyrite. $FeS_2$** A very common sulfide mineral in igneous, metamorphic and sedimentary rocks. It has a brassy yellow color with a bright metallic luster and cubic structure

**Pyrometric cone** A triangular pyramid piece of ceramic material designed to melt at a specific temperature. It is useful to measure and to control the heat in the kiln and to determinate when the firing is complete

**Pyrometric Cone Equivalent (PCE)** An index of refractoriness of a material. It is obtained by comparison of test cones of the sample with standard pyrometric cones

**Pyrophyllite. $Al_2Si_4O_{10}(OH)_2$** Hydrous aluminum phyllosilicate. It is the product of hydrothermal alteration of feldspars in Al-rich rocks

**Quartzitic sandstone** Silicilastic sedimentary rock composed of grains of sand (0.06–2 mm in diameter) that are predominantly quartz

**Refractoriness** Property of refractory materials to resist melting, softening or high-temperature deformation. Measured by determining the pyrometric cone equivalent (PCE)

**Refractory** Natural or artificial material, usually non-metallic with an unusually high melting point and that maintains its structural properties at very high temperatures

**Sandstone** Clastic sedimentary rocks formed by grit size clasts (between 1/16 and 4 mm). These grains are held together by matrix (small particles) or chemical cement, which in most cases is siliceous or calcareous.

**Secondary mineral** Mineral formed at the expense of primary ore 305 through processes such as weathering and hydrothermal alteration

**Sedimentary basin** A broad low area in the earth's crust in which sediments accumulated and lithified to form sedimentary rocks

**Sedimentary cycle** Succession of stages in a sedimentary basin leading to the formation of a sedimentary rock. The stages are due to tectonic and climatic changes and including the following processes: uplift, weathering, erosion, depositional, lithification and diagenesis

**Sericite** Fine-grained variety of muscovite

**Shale** Fine-grained sedimentary rock composed of silt and clay-size particles. Shale is distinguished from other mudstones because it is fissile, which means that tends to split into thin layers

**Slickensides** Pedogenic slickensides are convex-concave slip surfaces that form during expansion/contraction in expansive clay soils (Gray and Nickelsen 1989) when swelling pressures exceed shear strength at depths where vertical movement is confined and result in the development of structures "gilgai" (Watts 1977). These types of slickensides are oriented randomly, in contrast to tectonic ones which generally are oriented in response to structural deformation (Driese and Foreman 1992)

**Siliciclastic** Sedimentary rock formed by clastic or detrital material of siliceous composition

**Siltstone** Fine-grained sedimentary rock composed mainly of silt-sized particles (between 1/16 and 1/256 mm)

**Smectite** A group name for 2:1 layer phyllosilicate clay minerals. They are hydrate aluminosilicate with high capacity to absorb water between their layers (swelling clay minerals). Smectite minerals have large specific surface areas as well as large cation exchange capacity

**Sparite** 1 Crystalline carbonatic cement, with individuals whose size is above 10 microns (usually between 20 and 100 microns). 2 Limestone in which the fundamental component is chemically precipitated sparite

**Sparitic (See sparite)** Characteristic of crystalline cement

**Stratigraphy** Branch of geology that deals with the study of rock successions and the correlation of geological events and processes in time and space

**Stratotype** An interval of a stratigraphic section that constitutes the model to define and recognize a stratigraphic unit or boundary between two units

**Stratum** A layer of sedimentary rock of essentially homogeneous composition bounded by surfaces called bedding planes, which represent changes in the conditions of sedimentation

**Stylolite** Zig-zag structure that develops in the rocks during diagenesis, by pressure-dissolution processes. Stylolite amplitude is a measure of the minimum amount of soluble mineral (carbonate, silica, etc.) removed during formation

**Tectonic or orogenic cycle** Sequence of events leading to formation and then to the destruction of a mountain belt

**Telogenesis** It is a stage of diagenesis. According to Worden and Burley (2003) in the telogenetic regime the waters associated with early diagenetic processes are displaced by subsequent influx of meteoric water. The inward flow of rain water tank must be driven to it by a pressure head associated with rainfall areas. Basin inversion and associated fault movement, uplift, erosion and the formation of mountain exposed to meteoric water influx lead to significant changes in the types of basic-scale fluid flow

**Tonalite** Intrusive igneous rock, with phaneritic texture. It has a felsic composition represented by quartz (>20 %) and plagioclase feldspar (oligoclase or andesine), mafic minerals and minor alkali feldspar (<10 %)

**Transgression** Landward migration of the shoreline. It is caused by a relative sea-level rise

**Twin** Is a symmetrical intergrowth of two or more single crystals of the same substance. The twinned crystals are related by some simple symmetry operation (rotation or reflection)

**"Varied clays"** They have not a definite mineralogical composition. They may contain illite, smectite and/or chlorite, besides varying proportions of regular or irregular interstratified minerals, and if there is kaolinite, it is in low proportion, which makes their behavior in industry very difficult to predict. "Varied clays" have no defined application, because their specific use depends on technological tests. All "varied clays" of the province of Buenos Aires are sedimentary and the most important deposits are in Olavarria and Barker (Olavarria, Las Aguilas and Cerro Negro formations)

**Vermiculite. $(Mg,Fe,Al)_3((Al,Si)_4O_{10})(OH)_2\cdot 4H_2O$** Is a high-charge 2:1 phyllosilicate clay mineral. It is generally a product of weathering of micas. The name derives from Latin "vermicularis," because upon intense heating crystals become elongated, twisted and curved. Vermiculite has a special property of expanding after heating called exfoliation. It usually expands between 20 and 30 times its original size. Vermiculite has low thermal capacity and high thermal insulation properties

**Weathering** Is the disintegration or decomposition of rocks, through the physical and chemical action of atmospheric agents

**XPL** Cross polarized light

**Zircon. $ZrSiO_4$** Accessory mineral found in igneous rocks and some metamorphic rocks. Due to its chemical stability it is found as alluvial grains in some sedimentary deposits. Zircon contains trace amounts of uranium so it is very useful for isotopic data

# References

Brewer R (1960) Cutans: their definition, recognition and interpretation. Eur J Soil Sci 11(2):280–292

Driese SG, Foreman JL (1992) Paleopedology and paleoclimatic implications of late ordovician vertic paleosols, southern Appalachians. J Sed Petrol 62:71–83

Gray MB, Nickelsen RP (1989) Pedogenic slickensides, indicators of strain deformation processes in red bed sequences of the Appalachian foreland. Geology 17:72–75

Holtzapffel T (1985) Les mineraux argileux. Societe Geologique du Nord Publication, vol 12. p 136

Keller WD (1968) Flint clays and flint facies. Clay Miner 16:113–128

Watts NL (1977) Pseudo-anticlines and other structures in some calcretes of Botswana and South Africa. Earth Surf Process 2:63–74

Worden RH, Burley SD (2003) Sandstone diagenesis: the evolution of sand to stone. In: Burley SD, Worden RH (eds.) Sandstone diagenesis: recent and ancient, Reprint series of the international association of sedimentologists, Blackwell Publishing Ltd., New York, pp 3–44

# Index

**A**

Africa
 continent, 3
 South Africa, 1, 3, 6
  Karoo basin, 6
 South African Agulhas plateau, 6
Algae
 blue-green, 59, 63, 72, 151
 cellular colonies, 63, 100
 filament, 72
Algal
 mats, 8, 71, 107, 109, 111, 140, 151
Anatase, 19–21, 23, 30, 56
APS minerals
 alunite, 5, 27, 42, 43, 45, 74–76, 80, 81–85,
  147–148
 Ce-florencite-Svanvergite, 5, 80, 82, 84–85
Argillization, 133–135, 142

**B**

Bacteria
 *cyanobacteria*, 59, 64, 66, 72, 151
 *trichome*, 72
Basement
 argillized, 27, 29, 97
 bedrock, 16, 22, 27, 29, 30, 32–34, 36, 98,
  157
 Precambrian, 2, 3, 6, 9, 148, 152, 153, 158
 saprolite, 9, 15, 17–21, 23, 24, 27–32, 34,
  36, 37, 93, 95, 98, 100
 saprock, 15, 16, 18–24, 27–32, 34, 36, 37,
  46, 97, 98, 100
Basin (Tandilia)
 geodynamic evolution, 145, 146
 inversion, 85, 87, 120, 146
Bentonite, 123, 125, 129, 130, 152
Bioherms, 47
Biosignatures, 62, 64, 101
Biotite, 18–21, 29, 30, 33, 59, 155, 157

Bioturbation, 69, 152
Boundstones, 49, 57, 61, 70, 71, 72
Breccia
 chert, *see also* silicified, siliceous
  Las Aguilas Formation, 8, 74, 76, 80
  Cerro Largo Formation, 108
 hydraulic, 114
 limestone, 8, 43, 80, 125
Buenos Aires Province
 main producer of industrial rocks, 2
 Tandilia System, 1, 2, 5, 6, 62, 100, 101,
  107, 108, 114, 133, 137, 138, 148
 Ventania System, 5, 6, 86, 114, 147
Burial, 5, 60–62, 66, 69, 72, 85, 119, 121, 145,
 153
Burrows, 138, 139

**C**

Carbonate rocks
 biochemical
  boundstone, 49, 57, 61, 70–72
 clastic
  grainstone, 70, 76, 78, 79, 86
  mudstone, 8, 70–72, 74, 108, 119, 137,
   140, 151, 159
  packstone, 67
  wackestone, 67
 diagenetic
  dolostone, 47, 49, 50, 51, 53, 57, 59, 61,
   64, 68, 73, 154
 facies, 47, 49, 53, 71
Chlorite, 34, 44, 77, 152, 160
Clay coatings
 perpendicular
  diagenetic, 60
 tangential (epimatrix)
  detrital, 2, 21, 61, 118, 119, 140, 141
Complejo Buenos Aires
 basement rocks, *see also* bedrock

© Springer International Publishing Switzerland 2016
P.E. Zalba et al., *Gondwana Industrial Clays*,
Springer Earth System Sciences, DOI 10.1007/978-3-319-39457-2